CAMBRIDGE LIBRARY COLLECTION

Books of enduring scholarly value

Technology

The focus of this series is engineering, broadly construed. It covers technological innovation from a range of periods and cultures, but centres on the technological achievements of the industrial era in the West, particularly in the nineteenth century, as understood by their contemporaries. Infrastructure is one major focus, covering the building of railways and canals, bridges and tunnels, land drainage, the laying of submarine cables, and the construction of docks and lighthouses. Other key topics include developments in industrial and manufacturing fields such as mining technology, the production of iron and steel, the use of steam power, and chemical processes such as photography and textile dyes.

The Evolution of the Parsons Steam Turbine

Responsible for the generation of most of the world's electricity, and with applications to sea and land transport, the steam turbine may be regarded as a pivotal invention in the creation of a technologically advanced modern society. Charles Parsons (1854–1931) built the first practical steam turbine in 1884, and he remained at the forefront of its development for nearly fifty years, as he saw his invention become first the prime means by which thermal energy could be turned into electricity, and then the power behind pioneering cruise liners and warships. Alexander Richardson (1864–1928), an engineer and politician, had access to the inventor's papers when writing this account of the turbine's history. Published in 1911, and featuring more than 170 illustrative plates, it provides a valuable insight into the development of a technology that revolutionised power generation, marine transport and naval warfare.

Cambridge University Press has long been a pioneer in the reissuing of out-of-print titles from its own backlist, producing digital reprints of books that are still sought after by scholars and students but could not be reprinted economically using traditional technology. The Cambridge Library Collection extends this activity to a wider range of books which are still of importance to researchers and professionals, either for the source material they contain, or as landmarks in the history of their academic discipline.

Drawing from the world-renowned collections in the Cambridge University Library and other partner libraries, and guided by the advice of experts in each subject area, Cambridge University Press is using state-of-the-art scanning machines in its own Printing House to capture the content of each book selected for inclusion. The files are processed to give a consistently clear, crisp image, and the books finished to the high quality standard for which the Press is recognised around the world. The latest print-on-demand technology ensures that the books will remain available indefinitely, and that orders for single or multiple copies can quickly be supplied.

The Cambridge Library Collection brings back to life books of enduring scholarly value (including out-of-copyright works originally issued by other publishers) across a wide range of disciplines in the humanities and social sciences and in science and technology.

The Evolution
of the
Parsons Steam Turbine

*An Account of Experimental Research on the Theory,
Efficiency, and Mechanical Details of Land and Marine
Reaction and Impulse-Reaction Turbines*

ALEXANDER RICHARDSON

CAMBRIDGE
UNIVERSITY PRESS

CAMBRIDGE
UNIVERSITY PRESS

University Printing House, Cambridge, CB2 8BS, United Kingdom

Cambridge University Press is part of the University of Cambridge.

It furthers the University's mission by disseminating knowledge in the pursuit of
education, learning and research at the highest international levels of excellence.

www.cambridge.org
Information on this title: www.cambridge.org/9781108070089

© in this compilation Cambridge University Press 2014

This edition first published 1911
This digitally printed version 2014

ISBN 978-1-108-07008-9 Paperback

This book reproduces the text of the original edition. The content and language reflect
the beliefs, practices and terminology of their time, and have not been updated.

Cambridge University Press wishes to make clear that the book, unless originally published
by Cambridge, is not being republished by, in association or collaboration with,
or with the endorsement or approval of, the original publisher or its successors in title.

THE EVOLUTION

OF THE

PARSONS STEAM TURBINE.

Charles A. Parsons.

THE EVOLUTION

OF THE

PARSONS STEAM TURBINE.

AN ACCOUNT OF EXPERIMENTAL RESEARCH ON THE THEORY, EFFICIENCY, AND
MECHANICAL DETAILS OF LAND AND MARINE REACTION AND IMPULSE-
REACTION TURBINES; AND ON THE EFFICIENCY OF SCREW PROPELLERS,
ELECTRIC GENERATORS, AND MECHANICAL SPEED-REDUCTION GEARING
DRIVEN BY STEAM TURBINES;

A RECORD OF PROGRESS IN THE APPLICATIONS OF STEAM TURBINES IN
PROPELLING SHIPS, DRIVING ELECTRICAL MACHINES, AIR BLOWERS,
COMPRESSORS, FANS AND PUMPS, AND IN UTILISING EXHAUST STEAM
FROM OTHER ENGINES; AND

A DESCRIPTION OF THE MANUFACTURE OF TURBINES, AND OF THE WORKS
OF MESSRS. C. A. PARSONS AND COMPANY, HEATON, AND OF THE
PARSONS MARINE STEAM TURBINE COMPANY, LIMITED, WALLSEND-ON-TYNE.

BY

ALEX. RICHARDSON, A.I.N.A.

(With One Hundred and Seventy-Three Plates and numerous Illustrations in the Text.)

LONDON:
OFFICES OF "ENGINEERING," 35 AND 36, BEDFORD STREET, STRAND, W.C.

1911.

PREFATORY NOTE.

MUCH that appears in this volume has already been contributed by the Author to the columns of "ENGINEERING," notably the results of the application of the turbine for the propulsion of warships and merchant steamers, but the greater part is here published for the first time. This statement applies particularly to the review of the continuous experimental work which has been done for the purpose of improving the efficiency of the steam turbine for land and marine work, for evolving propellers well adapted to the higher rates of revolution desirable with turbine drive, for designing electric generators suitable for the high speeds necessary for economy, for the application of the turbine to air blowers, compressors, pumps, &c., and for the utilisation of the exhaust steam from reciprocating engines by turbines.

Obviously no such record would have been possible without access to the data derived from this experimental research, and the Author has much pleasure in expressing his great indebtedness to Mr. Parsons for placing the data at his disposal for the preparation of this book.

Thanks are also due to Mr. Parsons' colleagues—Sir William White, K.C.B., Mr. Gerald Stoney, Mr. Robert J. Walker, and Mr. H. Wheatley Ridsdale—for assistance in various directions, to Mr. William Menzies Johnston, the general manager of the Heaton Works, who has been identified with experiments and manufacturing there from the beginning, and to members of the staff at the various works—Mr. A. H. Law, Mr. C. C. Cook, Mr. A. Q. Carnegie, and others.

The Author desires also to acknowledge the help given by his colleagues, Mr. B. Alfred Raworth and Mr. H. Medway Martin, in the revision of the proofs, and by Mr. F. W. Jackson, who has been responsible for the preparation of the engravings.

LONDON, *March* 1911.

CONTENTS.

LIST OF ILLUSTRATIONS.

* See *Errata*, page xix.

ERRATA.

On page 164, on seventh line from bottom, "last" *should be* "first."

On Plate CXXXI., Title of Fig. 314 *should be* "Rotor of 6000-Kilowatt Alternator," *not* "Armature of 6000-Kilowatt Machine."

THE EVOLUTION OF THE
PARSONS STEAM TURBINE.

INTRODUCTORY AND HISTORICAL.

THE story of a great invention which has influenced the development of any department of thought or activity, and has aided the progress of the world, must, *primâ facie*, be of interest to the scientist and the sociologist, although to each the narrative appeals in different ways.

To the scientist it offers the promise of enlightenment upon allied problems, of suggestions for research upon lines diverging from the central idea, but yet depending upon it, and of encouragement to undertake the elucidation of subsidiary issues which may have been neglected by the inventor from lack of time or opportunity, or from the fact that their solution had not yet become pressing.

The sociologist finds in the history of invention the key to an understanding of many of the movements in the world which otherwise might perplex him. Indeed, it is impossible to explain many of the great political movements of the nineteenth century without a knowledge of the evolution of machines and mechanical processes, and of the reasons why certain inventions failed or languished for years while others had complete acceptance.

Again, the fascination and charm of inventions affect all who can appreciate the inspiring example of assiduous work undertaken primarily for the sake of science and for the satisfaction which comes from discovery, often carried out under difficulties and discouragement. To the true scientist material reward is only a secondary incentive; the establishment of truth and the dispersion of error are the guiding lights by which he directs his course, for it is only as he succeeds in approaching these that he can hope

B

to gain a reward for his exertions. Whether he is remunerated or not, science profits and the world's knowledge is enlarged.

The invention of the steam turbine by the Hon. Charles A. Parsons is a case which illustrates this view. In the first place, it can easily be proved that his invention has aided the progress of the world, since the turbine has cheapened mechanical power on land and sea, and has rendered possible steamship speeds greatly in excess of those attainable with the reciprocating engine. In the second place, these results have been achieved by a long process of theoretical and practical research directed to the elucidation of correct principles, whereby it became easy to eliminate the false or impossible hypotheses upon which the early workers proceeded without practical success, and to the evolution of mechanical appliances to conform to the physical laws established by experiment. In the third place, we have the inspiring example of an inventor searching assiduously for alternative methods when for a period of five years he was denied the right to proceed on the lines he originally devised and established by practical result as the most efficient, and in this search disclosing the falsity of hypotheses on which earlier workers attempted to succeed, and which are still accepted by some present-day inventors.

It is always difficult to trace the germ of such an invention. One thing is certain—Mr. Parsons did not find his first inspiration from a study of preceding turbine inventions. The effect of a knowledge of the failures of others might have been disheartening, for it must candidly be admitted that these early workers displayed great ingenuity and that their patent specifications contain suggestions of many fundamental ideas — the reaction principle, the divergent conical jet, the cupped bucket, and the principle of compounding by velocity or by pressure. And yet none succeeded. There are many explanations possible for such failure, amongst them an inadequate conception of the theoretical laws governing the expansion of steam, as well as difficulties of manufacture, the accurate workmanship necessary being inordinately expensive at that date.

Early in life Mr. Parsons developed an hereditary scientific and mechanical bent, and when at Cambridge, where he graduated eleventh wrangler in 1877, he made models of his epicycloidal engine. The first engine of this type was constructed when he was serving his four years' apprenticeship at Lord Armstrong's works at Elswick, while many others were manufactured during his two years' appointment (1881-1883) with Messrs. Kitson, of Leeds. This engine differed from all others of the period, for not only did the crank shaft revolve, but the cylinders rotated around it, making half as many revolutions as the shaft.[1]

We are not, however, concerned with this most ingenious invention, except for the fact that it led Mr. Parsons to tackle the problem of the steam turbine. He approached his subject from the hydraulic side. There was then little definite knowledge of the velocity of a steam jet, and it was considered by engineers generally that the construction of an efficient steam turbine was impossible.

Mr. Parsons undertook many experiments and finally adopted the idea of splitting up the fall in pressure over a great number of turbines in series, on the assumption that in each turbine the action would approximate to that in a turbine using an incompressible fluid such as water, and that the aggregate of these simple turbines, which together constituted the complete machine, would give the required efficiency approximating to that obtained in water turbines. This proved an accurate assumption; but a surface or bucket speed equal to a large fraction of the velocity of steam became a theoretical necessity. The mechanical difficulties of producing a turbine with a bucket speed approximating to 220 ft. per second were investigated and solved when Mr. Parsons became a junior partner and chief of the electrical department in Messrs. Clarke, Chapman, Parsons, and Company's works at Gateshead.

Before describing the mechanical features then devised, it may be well first to quote Mr. Parsons own definition of the three principal elements of the steam turbine.

" Firstly, in the steam turbine the working fluid is of low density, which implies a high velocity and necessitates the adoption

[1] See " ENGINEERING," vol. xxxix., pages 449 and 450.

of high surface speeds of blades, if only a few turbines are placed in series on the shaft. Where moderate surface speeds are essential, many turbines must be placed in series.

"Secondly, the working fluid is highly elastic and obeys the laws of adiabatic expansion of steam when giving out a large part of its energy in external work. This renders the use of diverging conical jets essential if the expansion is to be completed in one or in a few turbines. When there are many turbines in series, they must be a series of increasing capacity, to allow for expansion.

"Thirdly, although the working fluid may be dry, or superheated —and therefore a homogeneous gas—on entering the turbine, yet on account of expansion and the performance of external work, a portion of the fluid is condensed into minute drops of water, which are distributed among the remaining gaseous steam : and the working fluid thus becomes heterogeneous during the greater part of its passage through the turbine. This admixture of water with the gaseous steam has the effect not only of increasing the surface friction and the resistance in the steam passages, but also, when very high velocities of the working fluid are adopted, of cutting away the leading edges of the blades—an effect which becomes more pronounced the fewer turbines there are in series."[1]

This full enunciation of the principles of the steam turbine is the fruit of long experience and much research work, which will be reviewed in succeeding chapters. For the present the aim is to give the reader such an insight into the general subject as will enable him to follow the narrative of the development of the basal idea which we have described. The turbine, as we know it to-day, is in its essentials the same as the first machine made in 1884, but, in consequence of continuous experiment and of experience gained in actual practice, the details have been perfected and the efficiency greatly improved.

The steam turbine, as has already been indicated, comprises a series of rotating wheels, each complete in itself, and corresponding to a parallel-flow water turbine. Each wheel carries a ring of

[1] "Proceedings of the Institution of Civil Engineers," vol. clxiii. (Session 1905-06). Part I., page 171.

Fig. 1. The Rotor in the Lower Part of the Casing.

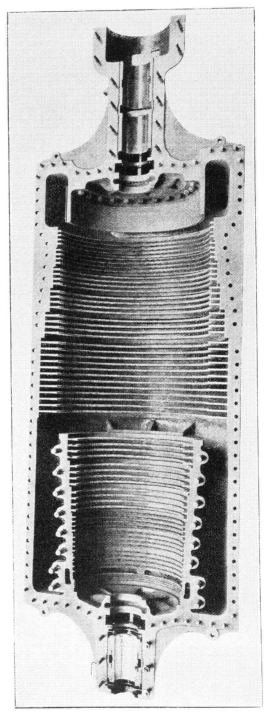

Fig. 2. The Lower Part of the Casing.

Plate I.—A Typical Turbine.

"Turbinia," built 1894 ; length, 100 ft. ; displacement, 44½ tons ; horse-power, 2300 ; speed, 32¾ knots.

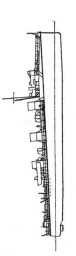

H. M. Torpedo-Boat Destroyer "Velox," built 1902 ; length, 210 ft. ; displacement, 420 tons ; shaft horse-power, 8000 ; speed, 27.12 knots.

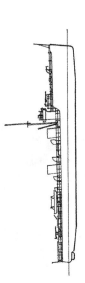

H. M. Torpedo-Boat Destroyer "Eden," built 1903 ; length, 220 ft. ; displacement, 540 tons ; shaft horse-power, 7000 ; speed, 26·22 knots.

H. M. Torpedo-Boat Destroyer "Swift," built 1908 ; length, 345 ft. ; displacement, 2170 tons ; shaft horse-power, 35,000 ; speed, 35.3 knots.

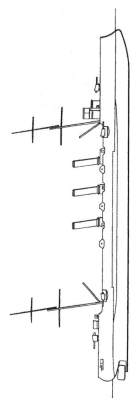

H. M. 3rd Class Cruiser "Amethyst," built 1905 ; length, 360 ft. ; displacement, 3000 tons ; shaft horse-power (estimated), 14,200 ; speed, 23.63 knots.

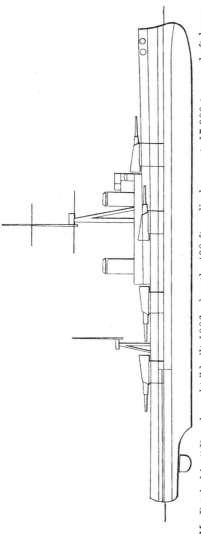

H. M. Battleship "Dreadnought," built 1906; length, 490 ft.; displacement, 17,900 tons; shaft horse-power, 24,712; speed, 21.25 knots.

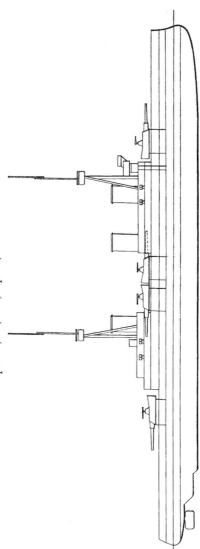

H. M. Armoured Cruiser "Invincible," built 1908; length, 530 ft.; displacement, 17,250 tons; shaft horse-power, 42,000; speed, 26 knots.

Plates II. and III.—Diagram showing Growth in Size between 1894 and 1908 of Turbine=Propelled Warships (Fig 3).

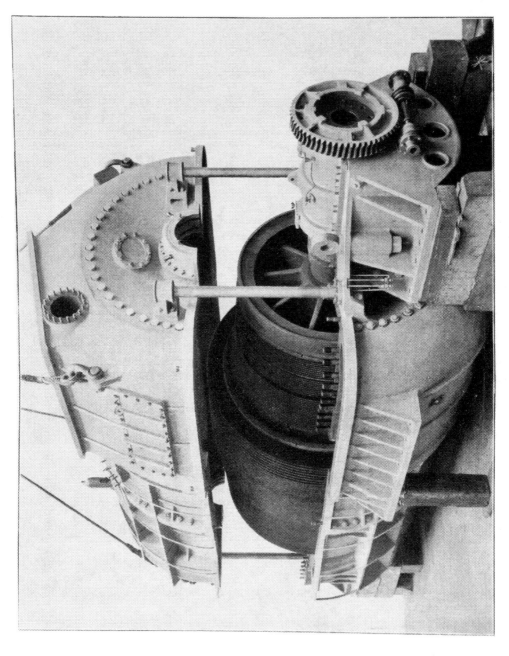

Plate IV.—A Typical Turbine, with Upper Part of Casing being Lowered into Position (Fig. 4).

blades, set and wedged with distance pieces, in a groove on a drum or rotor, and projecting radially outwards, as shown in the engraving on Plate I. In a cylinder or casing surrounding this rotor—and for that purpose cast in halves with a longitudinal flange connection —there are corresponding blades in grooves and projecting radially inwards, as illustrated in Fig. 2 on Plate I. The further view, on the Plate facing this page, illustrates another turbine with the upper half of the casing raised, showing that in the completed turbine the rings of rotor blades alternate with those in the stationary casing. The latter guide the steam to the former, and the blades are therefore set in opposite directions. The steam passes through the turbine in the annular space between the rotor and the casing and practically parallel to the axis, first entering a ring of fixed blades and impinging in a series of jets on the first ring of the moving blades, imparting rotating motion to the drum on which they are fixed. The steam issues from between these moving blades again in a series of jets, and the reaction of these jets further assists in the rotation of the drum.

The steam thus passes through each successive ring nearly parallel to the axis of rotation, preserving an almost constant longitudinal velocity, falling in pressure and expanding in volume. To allow for the increasing volume the angle of the blades was greater in each successive row in the earlier turbines, but in the later turbines the same result is achieved by increasing the length and therefore the area through the blades, except at the exhaust end, where blades are often of greater opening. For this purpose the drum and casing are increased in diameter in steps, forming expansion stages in the length of the turbine. In marine practice it is usual to divide the complete turbine into two parts—high and low pressure—in separate casings, and the drum is usually made parallel, with increasing length of blades in stages. Thus the velocity of flow is regulated so as to secure a high degree of efficiency. The blades range in length, in the case, for instance, of the "Mauretania's" and "Lusitania's" turbines, from $2\frac{3}{4}$ in. to 22 in., and in each complete installation there are 880,000 blades.

Many mechanical problems had to be solved in applying the

series of wheels or rings of blades in one turbine. The early turbines were all small with relatively slender drums, and, owing to the absence of precise workmanship and perfect material, vibration, with consequent heating of the bearings, was feared. Experiment enabled Mr. Parsons to produce a bearing permitting a little play. Governing gear, automatic lubricating devices, and other accessories had to be devised. The experimental work in arriving at satisfactory mechanism is reviewed in succeeding chapters.

The first turbine — a 6-electric-horse-power machine — was produced in 1884, and the driving of dynamos afforded a duty for which it was well adapted, especially when an electric generator suited to turbine drive was designed by Mr. Parsons. Other turbo-dynamos followed in quick succession, each marking an improvement on its predecessor as the result of close personal research work, and within five years there were made about 300 turbo-generators ranging in capacity up to 75 kilowatts. But in 1889 a dissolution of partnership deprived Mr. Parsons of his patent rights.

The next five years was a period of continuous work, under adverse conditions, at the works organised in 1889 at Heaton, Newcastle-on-Tyne, by Messrs. C. A. Parsons and Co. The parallel-flow system, which was the dominant idea of the original invention, could not be developed. But, undaunted by this hardship, Mr. Parsons set to work commercially manufacturing and experimenting with turbines of the radial-flow type, and although the results were largely negative, the record of valuable experiment, as told in a later chapter, is of great interest. As will be seen from the narrative, there was tested almost every previous turbine invention, and many more, which have since been claimed by patentees as new.

Several important improvements in detail, not only in turbine design but in the electrical machine, were introduced, so that when the patent rights were recovered in 1894 a great advance was possible. From this stage onwards progress has been continuous and rapid. The material and form of blades, and the method of building up the rings, formed the subject of extensive experiments; the system of governing passed through several stages towards thorough reliability; increasing dimensions, involving the use of stiffer drums, eliminated the tendency of the rotor to whip, and thus clearances could be reduced to a minimum between the tips of the blades of the rotor and the walls of the casing, and the blades on the casing and the surface of the rotor. The leakage of steam past the ends of the blades was therefore reduced, and the efficiency of the turbine consequently greatly increased. The cutting of the blade tips to a sharp edge was a great step in this improvement, as clearances of two hundredths of an inch (0.020 in.) became possible at the high-pressure end without liability to seizing or stripping of the blades. As turbines increased in size, too, the economy was greater as the ratio of clearance to blade area was reduced. Table I. shows not only the steady increase in the

TABLE I.—PERFORMANCE OF PARSONS TURBO-GENERATORS AT DIFFERENT EPOCHS.

Date.	Power.	Steam per Kilowatt Hour.	Vacuum. (Bar. 30".)	Superheat.	Steam Pressure per Square Inch.
	kilowatts	lb.	in.	Deg. Fahr.	lb.
1885	4	200	0*	0*	60
1888	75	55	0*	0*	100
1892	100	27.00	27	50	100
1900	1250	18.22	28.4	125	130
1902	3000	14.74	27	235	138
1907-10	5000	13.2	28.8	120	200

* These were non-condensing turbines using saturated steam.

size of turbo-generators, but the improving economy at different periods as a consequence of the developments described in succeeding chapters.

The first compound turbine was made in 1887, and the first condensing turbine in 1892, when there was at once a great advance in efficiency, the steam consumption being reduced from 55 lb. to 27 lb. per kilowatt-hour. By 1907 there was a further reduction to 13.2 lb. It was also established that, with the great range in expansion possible with compound turbines, the effect of an increase in vacuum of 1 in. at 26 in. was to decrease the steam consumption by 4 per cent., at 27 in. by 4½ per cent., at 28 in. by 5½ per cent., and between 28 in. and 29 in. by 6 per cent. to 7 per cent., a result facilitated by the introduction of the vacuum augmenter. Superheating was first applied in 1892, and it was found that each 10 deg. Fahr. of superheat reduced the steam consumption by 1 per cent.

In fifteen years electric station coal bills have been reduced to one half, and in twenty years to one quarter, and when one reflects upon the great extent to which electricity is used in industry, one can realise the immense social influence of the invention of the steam turbine. There are other directions in which economy has been achieved in factory operations. As narrated later, the low-pressure turbine and the vacuum augmenter render possible the utilisation of the exhaust steam from other engines for driving turbo-electric generators.

More phenomenal still has been the development of the turbine for the propulsion of ships. Moreover, this application of the system has appealed more directly to the public imagination because its results are more obvious. Only slight modifications were necessary in the turbine itself, and these will be described later, but great difficulties were involved in the adaptation of the screw propeller to the higher rate of revolution then necessary for turbine efficiency. The experimental investigation of this question was commenced about 1894, and has been pursued continuously up to the present time. To carry on this experimental work, a small company was formed in 1894, the Marine Steam Turbine Company, Limited, and it is only right that reference should be made here to those who co-operated financially with Mr. Parsons in the exploitation of the invention for marine purposes against a somewhat discouraging hesitancy on the part of shipowners. The directors included the late Right Hon. the Earl of Rosse, the late Mr. N. G. Clayton, Chesters, Northumberland; Mr. Christopher Leyland, Haggerston Castle, Beal, Northumberland; Mr. J. B. Simpson, now of Bradley Hall, Wylam-on-Tyne; and Mr. A. A. Campbell Swinton, 66, Victoria Street, Westminster; while the Managing Director was Mr. Parsons. Mr. H. C. Harvey, as Solicitor to the Company, took an active part in bringing about its formation. Subsequently, the late Mr. Norman C. Cookson, of Oakwood, Wylam-on-Tyne, joined the Board of the Company.

It is historically interesting to quote from the prospectus regarding the aims of the Company in view of the evidence which we give later as to the full realisation of these aims :—

"The object of the Company is to provide the necessary capital for efficiently and thoroughly testing the application of Mr. Parsons' well-known steam turbine to the propulsion of vessels. If successful, it is believed that the new system will revolutionise the present method of utilising steam as a motive power, and also that it will enable much higher rates of speed to be attained than has hitherto been possible with the fastest vessels.

"Up to within the last five years it has been found impracticable to obtain economical results from a motor of the steam turbine class, though such motors, on account of their light weight, small size, and reduced initial cost present great advantages over ordinary engines for certain classes of work.

"Recently—and more especially within the last two years—the steam turbine has been developed and improved. It has, further, been adapted for condensing, and results have been attained which place its performance as regards economy among the best recorded. Reports have been made upon it by the following well-known authorities :— Professor J. A. Ewing, F.R.S., Professor A. B. W. Kennedy, F.R.S., and Professor George Forbes, F.R.S.

"It is confidently anticipated that with turbines of, say, 1,000 horse-power and upwards, having a speed of revolution of about 2,000 per minute, the consumption of steam per effective horse-power will be less than with the best triple compound condensing engines.

"The initial cost of a steam turbine will be very considerably less than that of an ordinary marine engine of the same power. To these advantages must be added the consideration that the space occupied by the turbines will be very much less than that occupied by ordinary engines, thus largely increasing the carrying capacity of the vessel. The reduction in the amount of vibration admits of a diminution in the weight of the hull, which under the present system must be built stronger and heavier than will be necessary under the new system, in order to resist the effects of the vibration of the present class of marine engines.

"Another important feature is the reduced size and weight of the shaft and propeller. This will not only facilitate duplication and repair, and enable spare parts to be carried to an extent not hitherto practicable, but will also admit of screw-propelled vessels being used for navigating shallow waters, where at present only paddle steamers can be employed.

"The merits of the proposed system may be summarised thus :—Increased speed, increased carrying power of vessel, increased economy in steam consumption, reduced initial cost, reduced weight of machinery, reduced cost of attendance on machinery, diminished cost of upkeep of machinery, largely reduced vibration, and reduced size and weight of screw propeller and shafting.

"The efficiency of the screw propeller, the arrangements incidental to the adoption of higher speeds, the best form and the proportions and mounting of the propeller, the material of which it should be made, and the other points, can only be decided by investigation and practical experiment ; and it is to provide funds for the complete and exhaustive testing of the new system in these and other respects that the Company has been formed."

The first result of the formation of the Company was the laying down, in 1894, of the "Turbinia," fitted with the first Parsons marine turbine made. Three or four years of experimental work—described in a subsequent chapter—resulted in undoubted success. This was specially demonstrated at the Naval Review at Spithead in commemoration, in 1897, of the Diamond Jubilee of Queen Victoria's reign. The promise of extensive practical application justified the widening of the basis of the Pioneer Company and the formation of The Parsons Marine Steam Turbine Company, Limited, with a registered capital of £500,000, of which £244,000 has been issued and £211,200 called up. The Board of Directors was the same, with the exception of the late Mr. Clayton.

The first order for a turbine vessel was given by the British Admiralty in 1899 for the destroyer "Viper," and at the same time the Company contracted with Sir W. G. Armstrong, Whitworth, and Co., Limited, for the machinery of the destroyer "Cobra." This vessel was followed by the British Admiralty destroyers "Velox" and "Eden," and by His Majesty's cruiser "Amethyst" and other destroyers. France was the first of the Foreign Powers to fit the system to a warship: that was a torpedo boat built in 1903. The first turbine battleship—the "Dreadnought"—was ordered in 1905, and now nearly all British and foreign warships are fitted with the Parsons turbines. Thus, from the little 6-horse-power turbine made in 1884, we have in twenty-five years moved far beyond the range of the century of progress in piston engines. The latest marine installations are of more than ten thousand times this power, to give ships with the fighting capacity of battleships a legend speed which high authorities have declared would have been impossible without turbine machinery.

The diagram on Plates II. and III., adjoining pages 4 and 5, shows the various typical warships to which the turbine has been fitted, from the date of the "Turbinia" to those of the "Dreadnought" and "Invincible." These are all epoch-marking ships. The progress thus established, while thoroughly justified by the efficiency obtained on the trials of successive vessels, is also indicative of the courage and prescience of the technical advisers at the

Plate V.—The "Turbinia" alongside the "Mauretania" (Fig. 5).

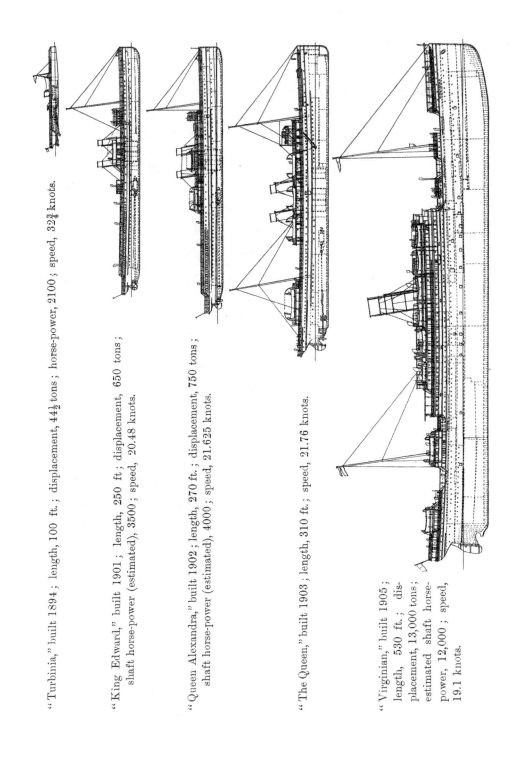

"Turbinia," built 1894; length, 100 ft.; displacement, $44\frac{1}{2}$ tons; horse-power, 2100; speed, $32\frac{3}{4}$ knots.

"King Edward," built 1901; length, 250 ft; displacement, 650 tons; shaft horse-power (estimated), 3500; speed, 20.48 knots.

"Queen Alexandra," built 1902; length, 270 ft.; displacement, 750 tons; shaft horse-power (estimated), 4000; speed, 21.625 knots.

"The Queen," built 1903; length, 310 ft.; speed, 21.76 knots.

"Virginian," built 1905; length, 530 ft.; displacement, 13,000 tons; estimated shaft horse-power, 12,000; speed, 19.1 knots.

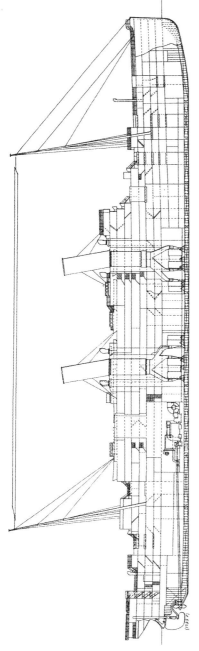

"Carmania," built 1905; length, 675 ft.; displacement, 30,000 tons; shaft horse-power, 21,000; speed, 20½ knots.

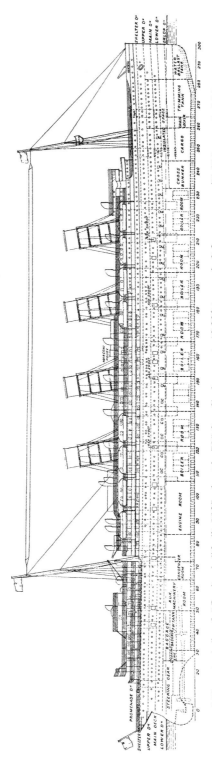

"Mauretania" and "Lusitania," built 1907; length, 785 ft.; displacement, 40,000 tons; shaft horse-power, 74,000; speed, 26 knots.

Plates VI. and VII.—Diagram showing Growth in Size of Turbine-Propelled Merchant Ships (Fig. 6).

T.S.Y. TURBINIA

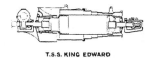

T.S.S. KING EDWARD

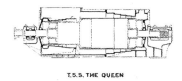

T.S.S. THE QUEEN

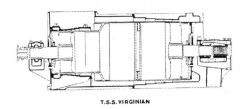

T.S.S. VIRGINIAN

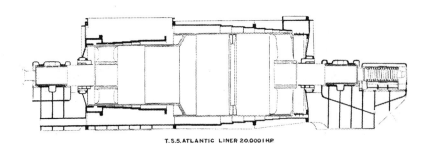

T.S.S. ATLANTIC LINER 20,000 I.H.P.

Plate VIII.—Diagram Showing Growth in Size of Marine Turbines (Fig. 7).

Admiralty. The first of the turbine-driven ships was ordered when Sir William White, K.C.B., was Director of Naval Construction and Engineer-Vice-Admiral Sir John Durston, K.C.B., Engineer-in-Chief of the Fleet. Their successors in these respective positions—Sir Philip Watts, K.C.B., and Engineer-Vice-Admiral Sir Henry Oram, K.C.B.—have shown corresponding scientific ability, and have assisted with equal zeal towards the development of the system and have, therefore, enabled speeds to be achieved which would have been impossible with reciprocating engines.

In the merchant marine the progress is equally significant. On Plates VI. and VII., adjoining this page, there is a diagram, illustrative of the growth in size of early merchant steamers fitted with turbines. It was not until seven years after the "Turbinia" was constructed that the first merchant steamer with turbine machinery was completed. This was the "King Edward," built in 1901; and while The Parsons Marine Steam Turbine Company had to become financially responsible to a degree, great credit is due to Captain John Williamson, of Glasgow, and Messrs. Denny, of Dumbarton, for their enterprise in becoming partners in the undertaking. This step was eminently justified in its result. A second steamer for the Clyde passenger traffic—the "Queen Alexandra"—followed in 1902. The first Dover and Calais mail turbine steamer—the "Queen"— was built in 1903, and several steamers were added for various services in 1904.

Then came the most important step in the development of the marine turbine : the appointment of a Commission of experts by the Cunard Company to investigate the suitability of the turbine for two proposed high-speed Atlantic liners, which ultimately became, as a consequence of the report made by the Commission, the turbine-driven 26-knot liners, the "Mauretania" and "Lusitania." Plate V., facing page 10, gives a view of the "Mauretania," with the "Turbinia" alongside.

Even before the Commission completed their investigations, the Allan Line decided to fit the system to two new steamers—the "Virginian" and "Victorian," and the Cunard Company adopted the system for the 19-knot steamer "Carmania." The Admiralty about this time decided to adopt the system for all warships. Thus there was no further question of the efficiency of the marine turbine, and practically every navy board in the world authorised its application to warships within a few months of the success of the "Dreadnought."

The Admiralty moved from the 10,000 horse-power of destroyers to the 15,000 horse-power of the "Amethyst," the 23,000 horse-power of the "Dreadnought," the 41,000 horse-power of the "Indomitable," and the still larger horse-power of the "Lion" and "Princess Royal"—a gradual rise, but still a great one, indicative of a progressive spirit at Whitehall. When the largest power in any completed turbine merchant ship was the 8000 horse-power in the Channel steamer "Queen," the decision to adopt the turbine for from 70,000 horse-power to 80,000 horse-power in Atlantic liners was still more eloquent of the enterprise of the Cunard Company and of the merchant ship-owning industry of Britain.

The diagram, on Plate VIII., facing page 11, illustrates the growth in the size of separate marine turbines. As shown, the largest turbine yet completed is the low-pressure engine of the "Carmania," designed to run at low revolving speed, and including the go-astern as well as the go-ahead machine ; in other large power marine installations—those, for instance, in the "Mauretania" and "Lusitania"—the astern turbines are separate.

The increase in economy is clearly established by the records, especially of the steam consumption of notable vessels given in Table II. on the opposite page, and more fully discussed in subsequent chapters. In view of these results, it is not surprising to note the rapid development in the application of turbines between 1906 and 1908, as indicated on the curve opposite. This rate of development has steadily increased.

Table III., on page 14, shows the power of the marine turbines fitted to the respective types and nationality of ships. It will be seen that the Parsons turbines built, or in process of building, at the end of June, 1910, represented collectively an aggregate of 4,700,250 shaft horse-power, a figure which also goes

TABLE II.—PERFORMANCE OF NOTABLE SHIPS WITH PARSONS TURBINES IN
DIFFERENT EPOCHS.

Date.	Name of Ship.	Length.	Displacement.	Horse-Power.	Steam Consumption per S.H.P. per Hour for all Purposes.	Speed in Knots.
		ft.	tons		lb.	knots
1897	" Turbinia "	100	44½	2,300	15	32.75
1901	" King Edward "	250	650	3,500	16	20.48
1905	" Amethyst "	360	3,000	14,200	13. 6	23.63
1906	" Dreadnought "	490	17,900	24,712	15.3	21.25
1907	" Mauretania " and " Lusitania "	785	40,000	74,000	14.4	26.0

to demonstrate the realisation of the aims set out in the prospectus of the Marine Steam Turbine Company already quoted.

Similarly satisfactory statistics might be published for the

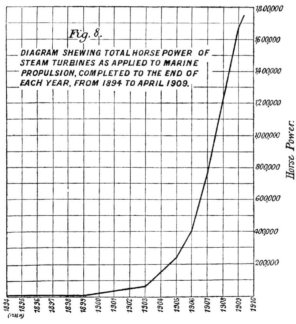

Fig. 8.

DIAGRAM SHEWING TOTAL HORSE POWER OF STEAM TURBINES AS APPLIED TO MARINE PROPULSION, COMPLETED TO THE END OF EACH YEAR, FROM 1894 TO APRIL 1909.

land turbine; but enough has been written on this phase of the subject. Before departing from it opportunity may be taken of mentioning those who were prominently associated in the pioneer work. First, at the Gateshead Works, there were the late Mr. Thomas Alderton, who was in charge of the electrical department, Mr. J. B. Willis, who was head draughtsman, and Mr. Richard Williams, who was assistant works manager. These associates of Mr. Parsons joined his staff at the Heaton Works when they were

TABLE III.—PARSONS TURBINES COMPLETED AND UNDER CONSTRUCTION, JUNE 30, 1910.

Country.	War Vessels.		Mercantile Vessels.		Yachts.		Total.	
	No.	Horse-Power.	No.	Horse-Power.	No.	Horse-Power.	No.	Horse-Power.
Great Britain and Colonies	143	2,116,800	56	533,900	6	17,000	205	2,667,700
United States	20	322,000	6	40,000	4	12,200	30	374,200
Germany ...	18	507,100	...	...	...	...	18	507,100
France... ...	16	235,750	4	62,000	...	...	20	297,750
Japan	5	93,100	7	76,000	...	...	12	169,100
Austria ...	3	70,000	...	...	...	...	3	70,000
Belgium ...	...	...	3	38,000	...	...	3	38,000
Italy	7	135,500	2	24,000	...	...	9	159,500
Russia... ...	4	168,000	...	...	...	...	4	168,000
Spain	16	102,500	...	...	...	...	16	102,500
China	1	6,000	...	...	...	...	1	6,000
Sweden ...	1	7,000	...	...	...	...	1	7,000
Denmark ...	2	8,000	...	...	...	...	2	8,000
Argentine ...	4	72,000	...	...	...	...	4	72,000
Brazil	3	38,400	...	...	...	...	3	38,400
Peru	...	...	2	15,000	...	...	2	15,000
—	243	3,882,150	80	788,900	10	29,200	333	4,700,250

commenced in 1889, as also did Mr. Gerald Stoney, who, although at the Gateshead Works, was not then so closely identified with the turbine as in after years, during which he has done invaluable work as a coadjutor of Mr. Parsons in his experimental research. When the first marine turbine was made, in 1894, the designs of hull and turbines respectively for the "Turbinia" were prepared by the late Mr. Robert

Barnard, and by Mr. J. B. Willis, and his assistant, the late Mr. Archibald Wass, who afterwards became head draughtsman at the marine works. Mr. Barnard, who was also in charge of the building of the "Turbinia," afterwards became manager at the marine works.

In order to explain generally the later and prospective developments of the Parsons systems it is necessary to indicate the main characteristics of other designs. As we have shown, the dominant principle of the Parsons turbine is the gradual expansion of the steam by small drops of pressure through each of several stages of the turbine with gradually increasing volumetric capacity. As this drop of pressure occurs both in the fixed and the moving rings of blades, the Parsons turbine is of the reaction type. There have since been introduced impulse turbines, with the whole expansion taking place between the fixed blades, but embodying the principle of expanding the steam in successive stages, but with larger drops of pressure. Amongst these are the Rateau and Zoelly turbines. Differing from these is the machine with expansion in a single stage through a divergent jet, as in the De Laval turbine.

There have since been developed turbines embodying the principle of velocity compounding. In these the drop of pressure is divided between from two to nine or more stages. The steam at each stage issues from a divergent nozzle, and, without change of pressure, passes in succession through a series of moving and fixed blades, which extract its kinetic energy, finally passing it on to another set of divergent nozzles, through which further expansion takes place. The kinetic energy of the steam resulting from this second expansion is extracted anew in a further series of moving and fixed blades. This type includes the Curtis turbine. In all impulse turbines the steam at the high-pressure end enters the moving blades over a small fraction of the total circumference of the wheel. This arrangement is known as working with "partial admission," and, as it is possible to vary the number of nozzles in use, it has been considered by some to give a reasonably high efficiency on board ship at low powers without the use of cruising turbines.

It has been claimed for the impulse marine turbine that it excels in efficiency at low powers, and although the claim has not been substantiated so far in practice, Mr. Parsons has recently been working at the problem, and has produced a partial-flow turbine which promises the same high economy at low, as at high, powers. This turbine is described in a later chapter. There has also been introduced, the practice of using one velocity compounding impulse stage at the high-pressure end with reaction blading for all subsequent stages.

In order that the economy of the turbine system may be further available for relatively low-speed steamers of between 12 and 18 knots, Mr. Parsons has devised a combination of piston engines and turbines for such vessels. Usually two piston engines are fitted to exhaust into one low-pressure turbine. The great range of expansion in a low-pressure turbine, in association with high vacua, enables it to develop a power equal to nearly 30 per cent. of the total when taking the exhaust steam from the piston engines at 8 lb. to 10 lb. absolute pressure, and expanding it to 1 lb. absolute. There are, as explained later, great potentialities for this system, the economy possible being from 12 to 15 per cent. This principle of using the exhaust steam from reciprocating engines is also extensively applied in iron and other manufactories where low-pressure turbines are used to drive electric generators. There is thus rendered available extensive motive-power and lighting current from the use of exhaust steam—formerly a waste product.

Successful results have also been achieved with turbines by interposing gear between the turbine and the propeller, so that the latter may run at a relatively much lower rate of revolution than the turbine. The Parsons Company are constructing two torpedo-boat destroyers in which the higher pressure portions of the turbine machinery, developing a large percentage of the total power, will be geared to the lower pressure portion of the installation. But these and other applications will be fully dealt with in later chapters. Here it may be said that the first application in a cargo steamship—the "Vespasian"—showed that, as compared with the triple-expansion engines first fitted, the turbines and gearing weighed 25 per cent. less and consumed 16 per cent. less fuel for the same power while steaming $9\frac{1}{2}$ knots, or, for the same fuel consumption, an increase of about 0.8 nautical mile per hour was obtained. Similar gear is being applied between the turbine and low-speed shafting, as, for instance, that for driving rolling mills. The great range between the speed of the electric generator and the rolling mill, both driven by turbines, proves the wide application of the system.

The success of the steam turbine has encouraged the engineering profession to investigate the problems of the internal combustion turbine which would dispense with boilers, and many auxiliary engines. One of the seemingly insurmountable difficulties is that the high temperatures of the gas would burn the turbine blades. It was suggested that this temperature might be reduced by the expansion of the flame in a divergent nozzle. Mr. Parsons and Mr. Stoney made some experiments with a flame flowing under pressure through such a divergent nozzle. A platinum wire was placed at each end, and the temperature effect on it observed through glass windows. It was found that the two wires attained almost identical temperatures, both showing a bright red heat. The gases had cooled on expansion, but the heat was restored by the impact of the gas against the wire. The same result would occur with blades, and thus there seems an insurmountable obstacle to the practical success of an internal combustion turbine.

But it has often been said that difficulties only encourage the engineer to greater effort. Hope for the future rests upon experience of the past, and the subsequent pages offer inspiring example, if not also helpful guidance, in the prosecution of experimental research, with all its seductions, beguiling at all times, if exasperating at some stages.

THE TURBINE FROM THE THEORETICAL STANDPOINT.

BEFORE entering upon a review of the experiments associated with the perfecting of the mechanical details of the turbine, it may be well to consider briefly the questions associated with the design of turbines from a theoretical standpoint. Experiments were made from time to time to ascertain the behaviour of the steam in its action on the blades and to acquire data for guidance towards higher efficiency. This work was carried on simultaneously with the experimental work described in the next chapter in connection with mechanical details.

Now, when the steam turbine has been extensively in use for many years, and can be designed with accuracy to fulfil the most varied requirements and to develop any power, it is difficult to appreciate the situation in 1884 when Mr. Parsons began to experiment and when there was a lack of the information and data since accumulated.

When the subject of the steam turbine was taken in hand, in 1884, very little was known as to the properties of steam in rapid motion beyond the fact that the resistance to the flow of steam in jets bore an approximate relationship to that of water at the same velocity for small amounts of expansion, and little as to the amount of the conversion of potential into kinetic energy when steam expanded through divergent nozzles.

It was thought better to proceed on comparatively sure ground and construct the first steam turbine on as close a parallel as possible with the water turbine, having regard to the different densities of the two working fluids. This line led directly to the splitting up of the expansion of the steam over a large number of stages, the action of the steam at each stage being closely analogous to that of an incompressible fluid or to that of water in the water turbine.

One great inducement to follow this course lay in the very high recorded efficiencies of water turbines, 75 per cent. to 80 per cent. being usually claimed in standard works. Another inducement was the reduction of the steam velocities, which not only permitted lower blade velocities but also obviated any risk of cutting the blades by the steam. The theoretical treatment was therefore developed as closely as possible from the simple hydraulic principles applying to the flow of an incompressible fluid, and attention was concentrated on the results obtained with small drops in pressure in a few simple turbine elements. For such small drops of pressure the working fluid might be regarded as incompressible. Experiments were also made with combinations of a greater number of such elements to form a complete engine.

Whilst the fluid might be treated as incompressible during the small expansion which takes place in a single element, this was not the case for the combination of a number of these in series where the increase in volume has to be considered.

The ratios of blade to fluid velocities found best in water turbines were as far as possible adopted. In view of the small variation in the efficiency caused by considerable divergence from the best velocity ratio — it was seen that the above simple method of design satisfied all practical requirements at that time. Use was made of the existing standard tables for the volume, pressure, and density of steam, and the hydraulic heads and velocities of issue corresponding to given percentage increments of volume were tabulated.

TABLE IV.—TYPICAL EXAMPLE OF EXPANSIONS, VELOCITIES OF ISSUE, AND DOUBLE VOLUMES.

Expansion per pair of rows	.015	.02	.03	.04	.05	.06	.07	.08	.09	.1	.16	.24
Velocity of issue (feet per second)...	164	189	231	266	296	324	349	372	394	415	516	622
Number of pairs required for a duplicate expansion	46.5	35	23.4	17.7	14.2	11.9	10.2	9.0	8.05	7.27	4.67	3.22

Table IV. is an example taken from Mr. Parsons' early notes on the parallel-flow turbine, calculated for steam in the neighbourhood of atmospheric pressure, for which approximately $pv = 56,000$ ft., p being absolute pressure in pounds per square foot, and v the specific volume in cubic feet per pound. The figures in the third line give the number of pairs of rows of blades necessary for a two-fold expansion of the steam when the ratio of expansion per pair of rows and the velocity of issue had the corresponding values given in the Table.

The Table gives sufficient data upon which to base a design for a given blade speed, the velocity of the steam being chosen to bear its proper relation to that of the blades, and the number of pairs of rows it was necessary to provide, for an expansion ratio of two, being thus obtained from the Table.

From beginning to end of such a turbine the blade height would obviously require to gradually increase, rising finally to twice its initial value. For practical convenience it was, however, sometimes made constant throughout, with a slight variation of blade angle or opening, and sometimes made to increase in two steps, the latter practice being generally followed in the Parsons turbine at the present day.

The experiments on the forms of blades to determine the best shape for maximum efficiency—described in the next chapter—led finally to the adoption in 1887 of a curved blade, the leading edge being nearly perpendicular to the direction of motion. In 1896 the blade form was further improved by a thickening of the back near the leading edge and a fining of the after edge. It is interesting to note that series of careful experiments made by the Parsons Company in recent years, comparing this type of blade with numerous other shapes, have not disclosed anything appreciably better than those adopted in 1896, which remain up to the present time the standard normal type of blading in the Parsons turbine.

The extension to the steam turbine of the hydraulic principles which apply to the conversion of the pressure energy of an incompressible fluid into velocity energy has the advantage that, in dealing in detail with the numerous elementary actions of which a complete turbine is built up, it considers a definitely measurable quantity, namely, the pressure drop, instead of, as in the heat drop method, the small difference between the two large quantities.

Of late years it has become necessary to go to greater refinements of calculation to obtain accurate estimates of power and steam consumption, and to design the blading to meet definite requirements with certainty. It is interesting, however, to show the connection between the methods at present adopted and the earlier practice, which still has many advantages, especially for experimental work. In the hydraulic method, the maximum energy per pound to be obtained from a given fall of pressure of an incompressible fluid, or one in which, for the small drop considered, the variation in density is negligible, is immediately expressed by the pressure head, and the corresponding velocity as the velocity due to this head under the action of gravity. In other words, if Δp be the drop in pressure in pounds per square foot for an incompressible fluid, or for any fluid whatever if such drop is small, the energy in foot-pounds per pound of substance, developed by expansion through the small fall of pressure in question, is expressed by $\dfrac{\Delta p}{w}$ where w is the density in pounds per cubic foot or by $v . \Delta p$ where v is the specific volume in cubic feet per pound. If V is the velocity corresponding to this head $V^2 = 2 \, g \, \dfrac{\Delta p}{w}$.

A complete turbine being made up of a number of such small elements, the complete energy is the sum of all the values of $v . \Delta p$, taking for v its appropriate value at each stage of the expansion.

This is perfectly general for any law of expansion. It can easily be shown that, for adiabatic expansion in particular, it gives identically the same value as the energy available from a pound of steam when made to undergo the series of processes constituting the cycle known as the Clausius, or Rankine, adiabatic cycle used as a theoretical basis in the design of reciprocating engines.

A further simplification results by introducing the well-known formula for isothermal expansion of a perfect gas, according to which the energy per pound obtainable from a substance expanding hyperbolically from pressure p_1 to pressure p_2 is expressed by

$pv.\log \frac{p_1}{p_2}$, pv having a constant value for the hyperbolic curve of expansion. When the curve of expansion no longer gives pv = constant, its variation, even for fairly large drops of pressure, is so small that an average value of pv introduced into the expression $pv.\log \frac{p_1}{p_2}$ still gives, with considerable accuracy, the energy per pound obtainable in the given expansion, and retains the advantage of depending on direct pressure measurements alone.

This expression also, when evaluated for a succession of small drops from a higher to a lower pressure, gives accurately the value of the energy obtainable from any fluid expanding along any curve between these pressures, and in particular when the expansion is adiabatic leads, as already stated above, to the same value for the energy as is given by the Clausius cycle.

EXPERIMENTS ON MECHANICAL DETAILS OF TURBINES.

MR. PARSONS constructed his first turbine in 1884-85, at the Works, at Gateshead, of Messrs. Clarke, Chapman, Parsons, and Company. For some years it was utilised on tests to determine the efficiency of alternative designs of mechanical details, and was ultimately presented to the South Kensington Museum. There it worthily takes a place among the great epoch-marking inventions which have enriched the world from an industrial standpoint.

This pioneer machine was described in Mr. Parsons first patents, filed on April 23rd, 1884. These were for "improvements in rotary motors actuated by elastic fluid pressure, and applicable also as pumps,"[1] and for "improvements in electric generators and in working them by fluid pressure."[2] Both patent specifications are lucid in their descriptions : they indicate a clear grasp of the mechanical necessities, and provide adequate measures for meeting them.

This first turbine was of the parallel-flow type, which time and experience have established to be in all respects the most efficient. The principle of expansion was enunciated clearly : the fluid operated in successive stages, the steam, it was stated, "undergoing expansion and falling in pressure in each until it leaves the last at a velocity not greatly above that which is practically attainable by the motor itself, although greatly above that practicable with a motor having oscillating or reciprocating parts." A feature introduced, but discarded later, except in engines of very high speed of revolution and of the exhaust type or for driving air compressors, was the use of what might be termed right and left-handed turbines "to balance the end pressure." The steam entered at the centre of the length of the turbine, and, dividing to right and left, exhausted at

[1] Patent No. 6735, A.D. 1884. [2] Patent No. 6734, A.D. 1884.

the ends. The blades were, of course, set at opposite angles in these respective halves so as to give a turning moment in the same direction.

There were also foreshadowed many developments which fuller knowledge has confirmed and more accurate workmanship has made practicable. For instance, there were introduced flexible bearings, which a long series of experiments later proved to be the only method of compensating for the vibration inevitable with a rigid shaft rotated at very high speeds, owing to the lack of minute precision inevitable in manufacture and the want of homogeneity of materials.

A complete description of this pioneer turbine is justified because of the thoroughness with which the design was worked out. The constructional details are clearly shown in the longitudinal section (Fig. 9) on Plate IX., facing this page, while perspective views of the turbine and electric generator and of the turbine alone are reproduced on Plate X., facing page 25.

The rotor was built up of rings of gun metal strung on the shaft, and held in place by screwed collars. The blades were cut at an angle of about 45 deg. out of the solid on the edges of the rings, as shown in Fig. 19 on Plate XII., facing page 30, but thus early it was announced that "in some cases it may be found convenient to make the blades of sheet metal, and to secure them in suitable grooves or recesses in the rings; in this way the blades may be made thinner, and they may be accurately formed, either before or after insertion in the rings." The blades, which varied in length and pitch according to the stages in expansion, were practically straight, and were sharpened on the edge at the entry. This method of cutting them straight and from the solid was adopted because of "existing facilities" of manufacture. It is stated, however, that "other forms of blades may be employed"; but it was not until considerably later that curved blades were actually introduced and found by experiments to give a higher efficiency.

The oil for lubricating the bearings was contained in a chamber at the end of the turbine, and was fed by a vertical stand-pipe; there were separate delivery and return pipes to and from all the

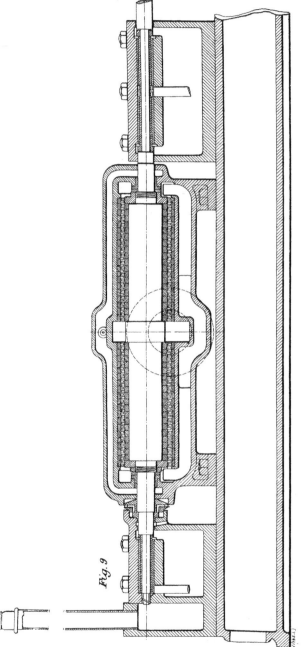

Fig. 9. Section of First Turbine made, 1884.

Fig. 9.

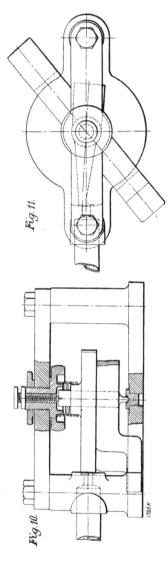

Fig. 11.

Fig. 10.

Figs. 10 and 11. Details of Electrical Governing Gear.

Plate IX.—The First Parsons Turbine, 1884.

Fig. 12. View of First Turbine and Generator.

Fig. 13. View of First Turbine.

Plate X.—The First Parsons Turbine, 1884.

bearings. Mounted between the turbine and the main bearing was a fan with a solid disc, having airways formed by drilling out radial holes connecting with transverse holes, as shown in the section, Fig. 9, to a cavity in direct communication, through a small pipe, with the top of the vertical stand-pipe. The working of the fan created a partial vacuum in the stand-pipe, owing to the centrifugal force acting upon the air in the radial holes in the disc. The partial vacuum primed a worm on the end of the turbine shaft, and this acted as a pump. Thus a constant circulation of oil was induced through the return pipes, bearings, and delivery main.

For the governing of the turbine there was fitted a leather diaphragm connected to the same fan, with a spiral spring to counteract the vacuum. This spring was loaded according to the speed of revolution regarded as the maximum ; in the event of this rate of rotation being exceeded, the excessive vacuum caused by the fan induced the diaphragm to overbalance the spring, and thus the throttle valve admitting steam to the turbine was regulated. The governor is illustrated in Figs. 12 and 13, on Plate X., facing this page, and the valve and liner in perspective on Fig. 18, Plate XI., facing page 26.

The speed of the turbine was further regulated by electrical connection with the dynamo to which it was coupled. The apparatus, shown in Figs. 10 and 11 on Plate IX., facing page 24, was mounted upon the field-magnets of the dynamo, and was operated by them. A bar of soft iron pivoted on the vertical spindle carried a pointer ending in a guard covering the inlet to the pipe at the left of Fig. 10, and communicating with the air-fan controlling the diaphragm. The polarity of the magnet tended to place this bar parallel to the lines of force—a tendency which was resisted by a spring attached at one end of the spindle, while at the other end was a bush for varying the tension of the spring. When the magnets were shunt-wound, and the iron bar was far from saturation, the intensity of the magnetisation varied as the electro-motive force at the terminals, and when series-wound it varied as the current. There was thus provided a twin-governor, the fan and diaphragm constituting a centrifugal governor, upon which was superimposed the fine regula-

E

tion of an electrical control governor. The supply to the air-pipe of the fan, and hence the vacuum behind the diaphragm, was regulated by the iron bar of the electrical governor, and thus the steam to the turbine was throttled, according to the extent of movement of the bar, combined with the suction of the fan.

The system of flexible bearings used was simple, and is illustrated in section and in perspective by Figs. 15 and 17 on Plate XI., facing this page. A light bush fitted on the journal formed the actual bearing. On this bush a series of rings or washers were strung, each alternate ring fitting the bush but not the casing, while those between fitted the casing but not the bush. Near the end there was on both the bush and casing a wide ring for centering the shaft, and a spiral spring, tightened up by a nut and washer, pressed all the rings together. Fig. 14 on Plate XI. shows an alternative flexible bearing that was tried but was not found as good as that of the washer type and Fig. 16 shows the ring for packing the gland against leakage of steam.

The steam seal at the end of the turbine shaft was formed of a metal washer fitting easily on the shaft at the end of the rings of blades, and abutting on a fixed bush, with passages beyond, connecting, through pipes, to a drain, where an injector was fitted to draw off the steam from the glands at the ends of the turbine, which was non-condensing.

The main claims of the pioneer patent[1] may be quoted, since they are of historical interest.

"(1.) In a motor the use in combination of a hollow cylinder with projecting rings of blades, and within it a solid or rotary cylinder with projecting rings of blades, upon which motive fluid is caused to act as it travels in directions parallel or approximately parallel to the axis of the solid or rotary cylinder, the arrangement and operation being substantially such as hereinbefore described with reference to the accompanying drawings.

"(2.) In a motor of the kind referred to in claim (1) the combination of two similar sets of rotary parts mounted upon one shaft, one set being placed at each side of the inlet for actuating fluid in such a way that the entering stream shall divide right and left and the exhaust take place at both ends, so as to balance or approximately balance the end pressures, substantially as described.

"(3.) In a motor wherein a moving cylinder or cylinders carrying blades on its

---

[1] Patent No. 6735, A.D. 1884.

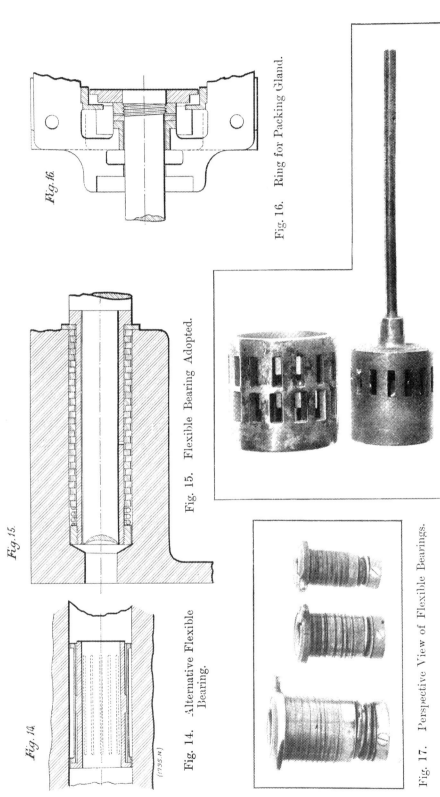

Fig. 14.

Fig. 15.

Fig. 16.

Fig. 14. Alternative Flexible Bearing.

Fig. 15. Flexible Bearing Adopted.

Fig. 16. Ring for Packing Gland.

Fig. 17. Perspective View of Flexible Bearings.

Fig. 18. Governor Valve and Liner.

(1735·N)

Plate XI.—Details of First Parsons Turbine Manufactured.

or their exterior is or are mounted on a shaft to rotate within a hollow cylinder or cylinders, also furnished with blades, mounting the said shaft in bearings having slight lateral play or elasticity combined with frictional resistance to such play in such a manner as to enable the moving cylinder or cylinders to rotate about its or their centre of gravity or principal axis instead of its or their geometrical centre or axis (if the centre of gravity and geometrical centre be nearly co-incident) and to cause the vibration to which the cylinder or cylinders may be subjected to be damped or modified, substantially as described.

"(4.) In a motor wherein a moving cylinder or cylinders carrying blades on its or their exterior is or are mounted on a shaft to rotate within a hollow cylinder or cylinders, also furnished with blades, the use for carrying the said shaft of elastic bearings, each comprising a bush and friction rings or washers pressed tightly together by a spring or springs in such a manner that the bush is capable of slight lateral movement resisted and controlled by the friction rings or washers, as described with reference to and illustrated in Fig. 14 of the drawings (Plate XI.).

"(5.) In a motor wherein a moving cylinder or cylinders carrying blades on its or their exterior is or are mounted on a shaft to rotate within a hollow cylinder or cylinders, also furnished with blades, the use of a centrifugal or screw pump mounted directly upon the motor shaft for forcing lubricant or cooling fluid to the parts to be lubricated or cooled, substantially as described.

"(6.) In a motor wherein a moving cylinder or cylinders carrying blades on its or their exterior is or are mounted on a shaft to rotate within a hollow cylinder or cylinders, also furnished with blades, wherein a pump that will not lift is employed to circulate lubricant or cooling fluid, the use of a suction fan to raise the level of such lubricant or cooling fluid in the return or suction pipe or chamber, and to enable the circulating pipe to start and to keep in action substantially as described.

"(7.) In a motor wherein a moving cylinder or cylinders carrying blades on its or their exterior is or are mounted on a shaft to rotate within a hollow cylinder or cylinders, also furnished with blades, the use for regulating the speed of the motor of a suction fan mounted on a motor shaft and a diaphragm connected to the spindle of the throttle valve, and caused by the action of the fan to operate in one direction in opposition to a spring in such a way as to open or close or alter the position of the throttle valve as described.

"Motors according to my invention are applicable to a variety of purposes, and if such an apparatus be driven it becomes a pump and can be used for actuating a fluid column or producing pressure in a fluid. Such a fluid pressure producer can be combined with a multiple motor according to my invention to obtain motive power from fuel or combustible gases of any kind. For this purpose I employ the pressure producer to force air or combustible gases into a furnace, into which there may or may not be introduced other fuel (liquid or solid). From the furnace the products of combustion can be led in a heated state to the multiple motor which they will actuate. Conveniently the pressure producer and multiple motor can be mounted on the same shaft, the former to be driven by the latter, but I do not confine myself to this arrangement of parts. In some cases I employ water or other fluid to cool the blades, either by conduction of heat through their roots or by other suitable arrangement to effect their protection."

Mr. Parsons primarily intended his turbine for driving electric dynamos and pumps, but contemplated in an additional claim " the use for pumping of two machines of the kind hereinabove described, coupled together so that one constitutes a motor and the other a pump, driven by such motor as hereinbefore described." This clause establishes the invention of the combination of the turbine and rotary air compressor afterwards brought into extensive use, and to be described in a later chapter. It also foreshadows the invention of the Armengaud and Lemale gas turbine.

Mr. Parsons indicated further that such a turbine might be applied to marine propulsion: " As the velocity is necessarily high, it will be advisable to place several fine-pitched screws on the shaft, in order to obtain a sufficient area of propeller blade ; and one screw may be prevented from interfering with another or others by suitable guide blades, or by other means." This established the claim for propelling ships by turbines, and enunciated the idea of multiple screw propellers, applied a few years later on the " Turbinia" and other fast steamers, as will presently be described, but subsequently discarded as experience showed that a single propeller on each shaft was more efficient.

In the same patent specification Mr. Parsons further says : " The motors, or successive portions of the compounding motor, may be arranged either upon one common shaft or upon different shafts. In the former case the first will deliver directly into the second, and the second into the third, and so on, the moving vanes of the second (say) rotating between its own fixed vanes and those of the third, and similarly for the others : the space for the actuating fluid increasing either continuously or step by step."

This, it will be seen, is a forecast of not only turbines in series, a system which has been universally applied in marine propulsion, but also of the tandem system of turbines, consisting of high-pressure and low-pressure turbines in separate casings on the compound principle, which, however, was not designed until some years after, and was first actually put into practice in the Elberfeld turbines.

The working of electric generators was the primary purpose for which the steam turbine was invented, and to this end Mr. Parsons designed and patented a dynamo to run at speeds up to 18,000 revolutions per minute.[1] Indeed, contemporaneous experimental work carried out on electric generators largely influenced the development of the steam turbine from this date onwards. For the present, however, it will be well to confine attention to the turbine, deferring for subsequent consideration the evolution of the turbo-generator and other applications.

Exhaustive trials were made with the turbine described, of 6 electric horse-power, coupled to the specially-designed electric generator. Experience first established the disadvantage of mounting the turbine and its bearings on separate pedestals on the bed-plates. Expansion troubles arose, owing to the turbine pedestals supporting the bearings becoming hotter than the pedestals, the immediate effect being to cause the blades to foul the casing. This difficulty was surmounted by casting the bearings as an integral part of the turbine casing, so that there was relatively more equal expansion and contraction. This procedure is still followed in land turbines, and the problem of keeping the rotor central, relative to the casing, is now largely a matter of good design and accurate workmanship, so that there is now no serious variation in its position owing to changes of temperature. Later, in large marine turbines the practice was introduced of bolting the bearings to the machine casing, the attachment, very rigid transversely, being placed as close as possible to the axis.

Experiments were undertaken to determine the best form of blade. This research was extensive, and was closely pursued at various periods in the development of the turbine, and here it may be appropriate to review the work done in connection with blading in the years between 1885 and 1888, before referring to the principal details in the turbines themselves. The main desideratum was to ensure the closest practical fit between the turbine casing and the blades of the drum and between the casing blades and rotor surface. A considerable clearance was necessary in practice, but leakage through the clearance space was a great cause of loss in efficiency. Although it was obvious that this loss would decrease rapidly as

[1] Patent No. 6734, A.D. 1884.

the size of the engine was increased, it was thought desirable to experiment with the view of reducing this loss in other ways. Various types of blading tried are illustrated on Plate XII., facing this page.

To take the place of rings with straight blades at an angle of 45 degs. cut out of the solid, as in the first turbine, rings for the rotor and casing were made with conical holes, and were tried in a turbine built especially for the tests. There was no gain in efficiency, and the manufacturing cost was greater.

In a turbine, which was made in 1885, for driving a dynamo at 10,000 revolutions per minute in the steamship "Earl Percy," there was fitted over the tops of the blades a shrouding, which was silver soldered; but, although this turbo-generator worked for many years, the shrouding led to no reduction in leakage and no improvement in efficiency. This type of blading is illustrated by the view, Fig. 20 on Plate XII., prepared from a photograph taken years afterwards and more clearly shown by the detail sketch reproduced on Plate XV., facing page 33.

Experiments were also made with sloped roots, as shown in Fig. 22 on Plate XII., so as to deflect the steam, and, by shielding the clearance space, to reduce leakage; but the difference was found to be inappreciable.

A notable feature in a 16-kilowatt turbine generator, made in 1887, was the introduction of baffle shrouding, as illustrated diagrammatically in Fig. 23, on Plate XII. To reduce leakage, the blades on the rotor were shrouded, and these, and the ring projections from the casing, worked in grooves in the casing and rotor respectively. The ribs projecting from the casing were made separately, and were calked in, and afterwards turned to fit the opposing ribs on the rotor, which were cut from the solid. This formation was costly to make, and the advantage gained in decreased leakage appeared to be somewhat counterbalanced by the loss due to increased skin friction. For this reason the baffle and channel were not at this time adopted except in the experimental engine.

Such shrouding, in a modified form, has been adopted in recent years in turbines manufactured by Messrs. Willans and Robinson,

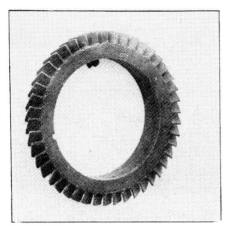

Fig. 19.   Straight Blading of First Turbine, 1884.

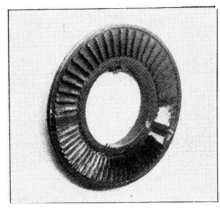

Fig. 20.   First Shrouded Blading, 1888.

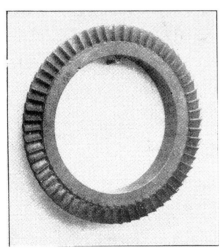

Fig. 21.   Curved Blades, 1888.

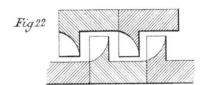

Fig. 22.   Blading with Sloped Roots, 1886.

Fig. 23.   Baffle Shrouding, 1887

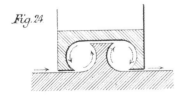

Fig. 24.   Vortex Flow type of Baffle Shrouding.

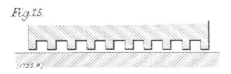

Fig. 25.   Serated Surface for Reducing Leakage.

Plate XII.—Blading and Dummies.

Plate XIII.—First Turbine with Stepped Cylinder for Successive Stages of Expansion (Fig. 26).

Limited, of Rugby, as will be described later. This baffle shrouding was found, in the 16-kilowatt turbine, to give no improvement in efficiency, the increased skin friction of the shrouding appearing to neutralise the reduction of leakage.

It was next thought that some form of vortex flow might be employed to reduce the discharge through an opening of given dimensions. A model in tin (Fig. 24) was made to try this formation. This model was tested with water at about 5 lbs. pressure, but the discharge appeared to be about the same as it would have been without vortex action, and the arrangement was therefore no better than the simple channel and baffle strip just described.

Experiments were simultaneously made by Mr. Gerald Stoney, who had been closely identified with Mr. Parsons in the work, to test the effect of serrating one surface and opposing this to a plane surface, as shown in Fig. 25, which, it will be seen, corresponds to ordinary water grooving. It was found that the discharge with water was the same as with plane surfaces the same distance apart and of the same total length. In other words, it was established that serrating one surface of a loose packing does not diminish leakage. Some years later Mr. Stoney tried the same experiment with steam at various pressures instead of water, and the same conclusion was arrived at. These early experiments also proved the great advantage of " labyrinth packing," now universally adopted, over any straight form of packing.

In a 32-kilowatt turbo-generator, made in 1888, the blades, still cut out of the solid, were bent so as to ensure a better shaped steam jet. This is illustrated by Fig. 21 on Plate XII. These curved blades were formed by undercutting the entry side of the ring, out of which they were machined, by a groove turned just below the roots, the blades being thereafter bent to the desired curve by pincers.

The turbine used for the tests of this last-mentioned blading was the first with a rotor constructed with steps of increasing diameter towards the low-pressure end, and the rings of blades were also increased in length. These changes were introduced in order to enable better expansion of steam to be obtained. This turbine[1] is illustrated in perspective on Plate XIII., facing the preceding page, and in the complete sectional drawings of the casing and rotor reproduced on Plate XIV., facing this page.

The several improvements, which are illustrated on the section, resulted in the steam consumption per kilowatt hour being reduced from 200 lb. in the old machines to 58 lb. in a non-condensing 32 kilowatt set, using saturated steam of 100 lb. pressure.[2] It was found that the curving of the leading edges of the blades improved the efficiency by about 25 per cent. over that with straight blades. The clearance had also been reduced to 0.015 in., which brought the average leakage down by from 15 per cent. to 20 per cent.

Many of these improved turbines were made, the most notable being several to drive at 5000 revolutions per minute 60-kilowatt dynamos, producing continuous current at 110 volts, and others to drive at 4800 revolutions per minute 75-kilowatt alternators. Of the latter, four typical sets were put to work at the station of the Newcastle and District Electric Lighting Company, and were the first turbo-generators used in a public electric supply station. Many sets were fitted for electrically lighting steam-ships, the largest up to this period being for the "City of Berlin." Their use for generating current to light electrically shipyards and other outside work—first tried at Elswick—proved most acceptable.

Research work was undertaken in 1888 with a new design of parallel-flow turbine, embodying the now generally-accepted principle of the dummy piston. This was introduced at one end to form a steam seal or check. The blades were only on one side of the steam inlet, and not on both as in the previous right and left-handed machines. This design was found to give a superior efficiency and to reduce the cost of manufacture. It has since been generally adopted, although the double-ended form is still used in some cases where the angular velocity is abnormally high, as in turbines driving air compressors, and also where very large volumes of steam have

---

[1] Patent No. 5312, A.D. 1887.
[2] See "ENGINEERING," vol. xlvi., page 478.

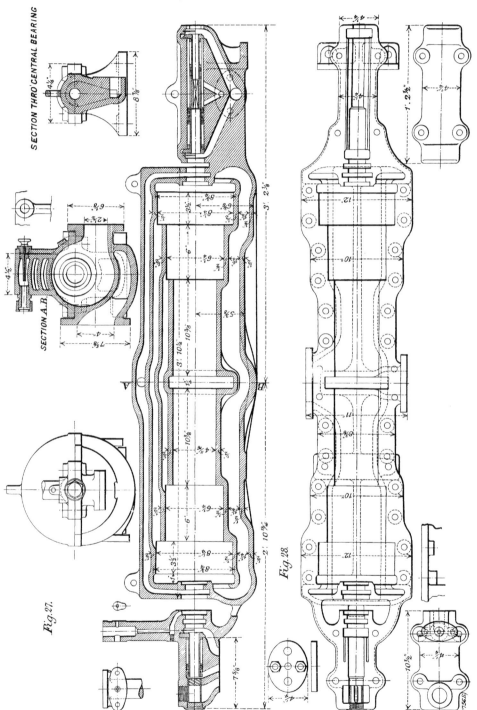

SECTION THRO' CENTRAL BEARING

SECTION A.B.

Fig. 27.

Fig. 28.

Plate XIV.—Section and Details of First Stepped Turbine, 1888 (Figs. 27 and 28).

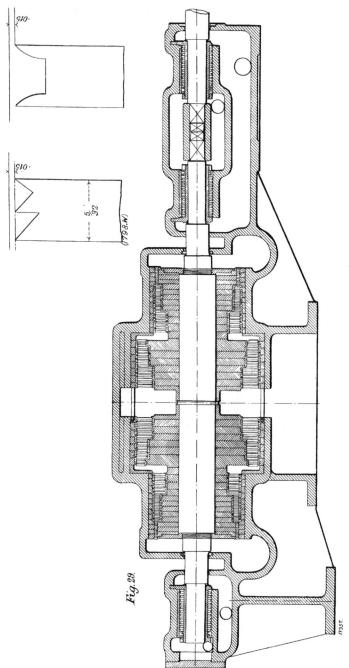

Fig. 29.

(798.H)

Plate XV.—First Low=Pressure Turbine, Designed 1889, with Shrouded Blading (shown in Perspective in Fig. 20, Plate XII.).

to be dealt with, as in exhaust turbines of large size and high speed.

By this time—four years after the first turbine was made—public interest in the invention had greatly increased, but few clients were found with sufficient enterprise to adopt the system for large powers. During the first five years of the invention, from 1884 to 1889, about 300 machines were manufactured, ranging up to 75 kilowatts. The steam consumption of 58 lb. per kilowatt hour then attained has not been much bettered in non-condensing turbines of this size, even up to the present day.

The progress in design during these first five years is clearly established by the drawings of a turbine designed early in 1889. This was a tandem compound non-condensing turbine, with a high-pressure and a low-pressure unit on the same shaft. The high-pressure machine corresponds to that illustrated in Figs. 27 and 28 on Plate XIV., and the low-pressure machine is shown in Fig. 29 on Plate XV. This turbine shows a rapid evolution in design as the result of painstaking and systematic experimental research.

The high-pressure turbine was that of the non-condensing 75 kilowatt turbo-generator set, constructed in 1888 for the Newcastle and District Electric Lighting Company, to supply single-phase current of 1000 volts at 80 periods, and running at 4800 revolutions per minute. It was of the double-ended type, with solid cut blades curved by bending them as already illustrated and described.

The low-pressure turbine, however, constituted the novel feature of this design for a compound machine. It was also of the double-ended type, with solid cut curved blades provided with a shrouding, as illustrated on the detail drawing on Plate XV., and already described on page 30. Steam was to be admitted at the ends at about atmospheric pressure, and to be exhausted into a condenser at the centre at about 1 lb. absolute. The joint at each end of the spindle, where it emerged from the turbine casing, was made steam tight by water admitted from the hot well or the feed tank, or by steam above atmospheric pressure passed from some part of the turbine to an annular groove round the spindle fitted with packing pieces, as in the earlier machines. This, in fact, is the first

F

invention of the steam-packed glands[1] now universally adopted in steam turbines.

This tandem compound turbine, however, was not constructed until twelve years later, when it became the historical Elberfeld machine, to be referred to in due course. The reason for this delay was a difference between Mr. Parsons and Messrs. Clarke, Chapman, and Company, which culminated in a termination of the partnership in 1889. The firm retained the rights in Mr. Parsons' patents, including that for the parallel-flow system as applied to steam turbines. Mr. Parsons guaranteed not to manufacture a parallel-flow turbine of any kind, nor to build a colourable imitation of his drum armatures, and for a time there was a check to the development of this fundamental, parallel-flow, principle, which experience had, even then, established as the most practical in turbine design.

The direct consequence was that Mr. Parsons and his colleagues had to apply their inventive ingenuity to the adaptation of mechanism for the utilisation of the radial flow of steam to turbine driving. The results, although largely negative, are still interesting. They establish the futility of many suggestions which even now recur in patent claims by other inventors.

The first of the experiments with the radial-flow principle were carried out with inward-flow jets, delivered from passages formed in a series of discs, made in halves and fitted into a casing which was divided longitudinally into upper and bottom halves. This arrangement is illustrated in Fig. 31 on Plate XVI., facing this page. A steam-tight joint was formed between these discs and the shaft of the rotor by means of white-metal rings cast into the discs. Into the periphery of the discs were fitted brass guide vanes, from which the steam played upon wheels of brass on the shaft of the rotor, having vanes cast on them like a water turbine. Indeed, the blading arrangement closely resembled that in Thomson's vortex turbine.

A radial-flow turbine with such discs and wheels was made in the Parsons works at Heaton, established in the same year (1890). This turbine became known as the "Jumbo,"[2] and was coupled direct

---

[1] Patent No. 5312, A.D. 1887.
[2] Patent No. 1120, A.D. 1890.

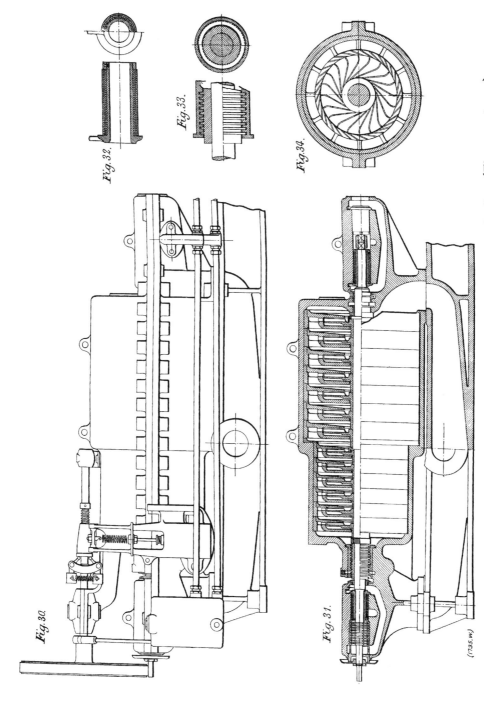

Fig. 32.

Fig. 33.

Fig. 34.

Fig. 30

Fig. 31.

(1735. w)

Plate XVI.—The Inward Radial-Flow Turbine, named "Jumbo," 1890 (Figs. 30 to 34).

Fig. 35. External View.

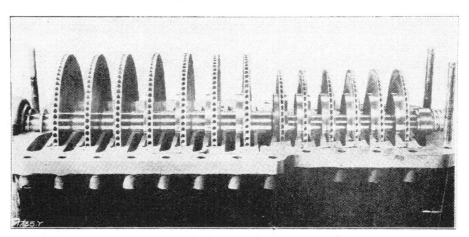

Fig. 36. Rotor with Wheels.

Fig. 37. Casing with Cells.

Plate XVII.—The Inward Radial=Flow Turbine, named "Jumbo," 1890.

to a 32-kilowatt generator, running at 6000 revolutions. It was afterwards utilised for a series of experiments to minimise steam leakage, which had continued in the earlier parallel-flow machines.

A section, and several important details, of the turbine "Jumbo" are given on Plate XVI. An external view of the turbine is given on Plate XVII., along with illustrations of the rotor, showing the wheels, and of the casing. The turbine consisted of a series of inward-flow discs and wheels, as already described, and well shown by the perspective views on Plate XVII., facing this page. The casing, made in halves with a longitudinal joint, carried the guide-vane discs, which were so arranged as to form a series of separate chambers, in each of which a brass wheel revolved. Each wheel thus rotated in a cell, and the steam, after driving the wheel, passed to again flow tangentially through the guide blades on to the vanes in the next wheel. Thus we have applied by Mr. Parsons, and enunciated in his patent of 1890, the principle of a cellular turbine for inward radial flow. This idea was not then pursued further because of the urgency of other problems. It was afterwards developed in parallel-flow turbines by Rateau, Zoelly, Curtis, and others.

The bearings of this turbine were flexible, but differed from those in the earlier machines. They are shown in Fig. 32 on Plate XVI., facing page 34. Over the bush there were three concentric tubes, held by a collar on the bush at one end, and by a nut at the other. A thin film of oil, fed through the holes, kept the bushes apart, and allowed such play of the shaft as might be caused by vibration, while the viscosity of the oil resisted powerfully any undue motion. This form of bearing was the outcome of experience with preceding machines, where a series of washers had been adopted. It was found that these washers cut into the bearing bush and into the casing around the bearing. The concentric tubes were found much more satisfactory, and the system is still adopted practically in all Parsons turbines running at speeds over 2000 revolutions per minute.

Another feature of this turbine was the adoption of a thrust block to sustain any want of end balance. (Fig. 33 on Plate XVI.)

In this machine, too, there was first introduced labyrinth packing as a steam seal or check as shown on Plate XVI. This consisted of projections on a bush over the shaft fitting into grooves on a block formed in two pieces to encase the shaft.

The speed-governing system adopted in the first turbine, and described on page 25, was modified, largely because the governing arrangement formed part of the patents retained by Messrs. Clarke, Chapman, and Company. As shown in Fig. 30 on Plate XVI., facing page 34, a projection of the shaft beyond the casing served to revolve by friction a wheel driving a shaft, which had a centrifugal spring governor controlling the position of a sleeve rotating with the shaft. The sleeve was parallel for part of its length, and cut away for the remainder, thus forming a cam. At every rotation this cam moved the spindle of the steam-inlet valve, overbalancing the spring momentarily. Thus steam was cut off a number of times per minute for periods which varied with the position of the cam as it was caused to slide one way or the other by the action of the governor, and this arrangement has since been known as "gust" governing. An electrical system involving the use of a solenoid was also adopted for the operation of the valve spindle. The development of the system of gust governing will be fully reviewed in connection with tests made later.

This inward-flow turbine proved a failure. It was found that, in the very short space of an hour, the guide vanes were destroyed, owing to "foreign" matter and water, imprisoned in each wheel case, being thrown alternately inwards by the steam and outwards by centrifugal force. New cast-iron wheels were made, and the guide-passages were formed by drilling holes in a tangential direction. These stood all right during the trial tests, but the resulting efficiency was not satisfactory. It was, moreover, found very difficult to drain the wheel chambers satisfactorily. With highly-superheated steam the turbine might have worked well; but even with it there is doubt as to whether the radial-flow principle would have proved comparable with the parallel-flow system.

Mr. Parsons at once discarded the idea of an inward-flow machine, and utilised the mechanism and casing of the "Jumbo" turbine for experiments with outward-flow wheels, with one row of blades. These gave a much better efficiency than the inward-flow

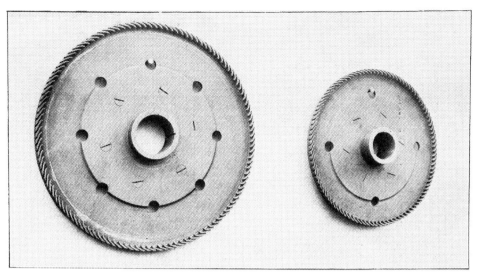

Fig. 38. Outward-Flow Wheel and Guide, with Single Row of Blades Tried in "Jumbo."

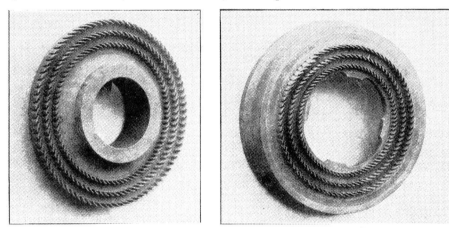

Fig. 39. Outward-Flow Wheel and Guide, with Three Rows of Blades, as used in Scotland Yard Turbine, 1890. (See Figs. 41 to 43.)

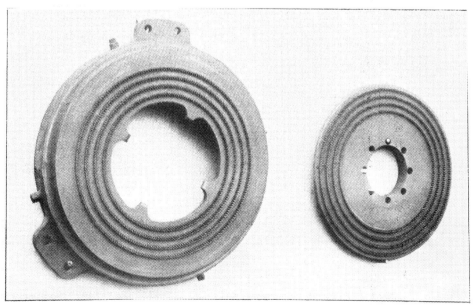

Fig. 40. Outward-Flow Wheel and Guide, with Four Rows of Blades, as used in Turbine Tested by Professor Ewing. (See Figs. 48 and 55.)

Plate XVIII.—Wheels and Guides for Outward Radial-Flow Turbines tried in 1890.

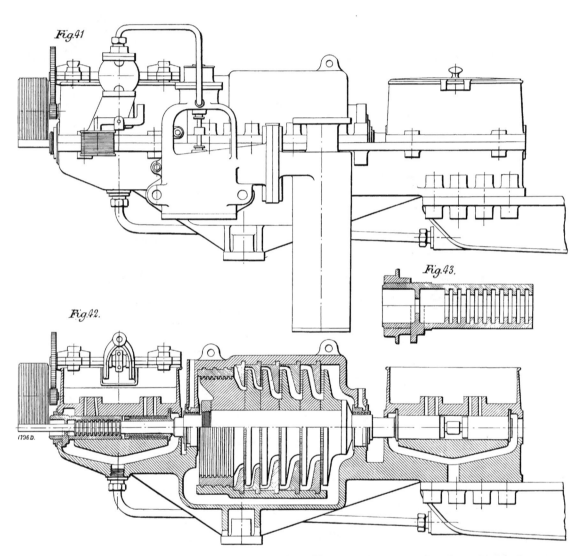

Fig.41

Fig.42

Fig.43.

1736.D.

Plate XIX.—32=Kilowatt Outward Radial=Flow Turbine for New Scotland Yard, London
(Figs. 41 to 43).

system. The outcome was the construction of the turbine known in the works as the " Mongrel"; it embodied amongst other improvements several rings of blades on the discs, as shown in the perspective views of guide and wheels (Fig. 39 on Plate XVIII). Several machines of this design were made, and amongst these were two for driving generators for lighting the New Scotland Yard Offices of the Metropolitan Police. These turbines are illustrated by the elevation and section on Plate XIX., facing this page. A perspective view is given on Plate XX., facing page 38.

It will be noted that this was the first of the radial-flow turbines with moving and fixed blades alternating and overlapping. The wheels were of brass with blades cut out of the solid. There were three or four rows of blades, and the blades were increased in size

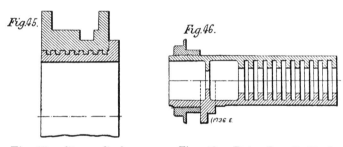

Fig. 45. Steam Seal.   Fig. 46. Rotor Longitudinal
Adjustment System.

towards the low-pressure end to allow for expansion and the drop in pressure. The fixed or guide blades were also formed out of the solid on rings or plates let into grooves in the casing. The steam was admitted behind the first of the rings with the fixed blades, and passed outwards between the fixed and moving blades, imparting turning moment, and returned behind the second ring, actuating in the same way the second wheel, and so on to the exhaust end. The steam spaces are shown in white on the section. The steam seal at the high-pressure end was formed of a dummy piston, with rings on the periphery engaging in grooves on the casing, and is shown in Fig. 45, above.

There was also introduced a system of longitudinal adjustment of the rotor. This was arranged at the thrust block, which was of the collar type, as shown in Fig. 46, above. The piston, or "dummy," for the steam seal was proportioned relatively to the wheels, so as to balance the spindle in the longitudinal direction as accurately as possible. The thrust block was in two halves, the upper sliding over the lower, which was bedded in the casing. Both halves protruded, and over the end a screw nut was put on, as shown in Fig. 46. Only the upper half of the block was screwed; the lower half was turned down to clear the threading of the nut, but was formed with a flange which abutted against the casing, while the nut had a flange also to abut against the casing. The turning of the nut thus caused the upper half of the thrust block to slide, bringing the spindle back, and reducing the clearance between the blades and their approximate surfaces, while the bottom half acted in compression, and prevented the moving blades coming into contact with the casing. At the same time the rings of the dummy were brought back against the rear surface of the grooves, lessening the possible leakage at this point also. The steam glands were also tightened by the same means. Adjustment similar to this in principle still continues in use in all Parsons turbines.

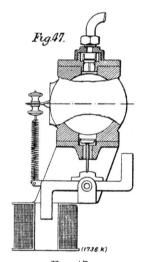

Fig. 47.

FIG. 47.
ELECTRICAL GOVERNOR.

The mechanical governing system on former turbines was only slightly modified, as shown on the elevation, Fig. 47, annexed. A friction wheel, driven from the end of the spindle, rotated a countershaft, on which was an air pump controlling the steam valve in a casing mounted alongside the turbine on the same bedplate. At the bottom end of the valve spindle a connection was formed with a lever actuated by a solenoid controlled from the dynamo driven by the turbine. Geared wheels were used afterwards, instead of friction wheels.

This type of turbine gave encouraging results. In addition to the one 16-kilowatt and two 32-kilowatt turbines for lighting the headquarters of the Metropolitan Police, two 75-kilowatt machines were supplied to the Newcastle and District Electric Light Company,

Plate XX.—32-Kilowatt Outward=Flow Turbine for New Scotland Yard, London (Fig. 44).

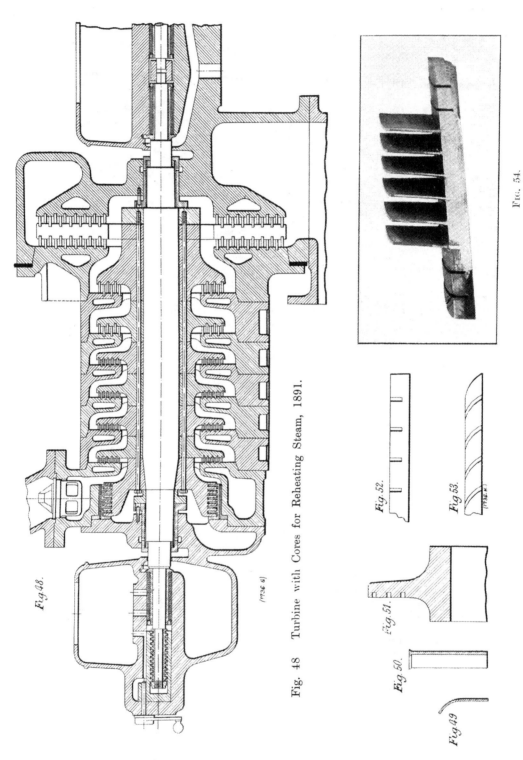

Fig. 48.   Turbine with Cores for Reheating Steam, 1891.

(1736 s)

Fig. 52.

Fig. 51.

Fig. 50.

Fig 53.

Fig 49

Fig. 54.

Figs. 49 to 54.   Details of Separately Secured Blades.

Plate XXI.—Outward Radial-Flow Turbine, with Reheating Cores.

Limited. This design of turbine, however, proved difficult to manufacture and was expensive. Moreover, the steam efficiency was not considered entirely satisfactory. It is noteworthy that one of the 32-kilowatt sets is still at work supplying electricity for lighting and running machinery at Scotland Yard.

Experience corroborated the experiments already referred to, that there was considerable loss of power owing to the presence of water in the steam as the result of boiler priming and steam expansion in the turbine. The idea of using superheated steam was considered, but at the time a system of jackets for reheating the steam was preferred.[1] The steam was passed through spaces cored in the diaphragms projecting from the casing inwards to carry the guide blades. These cores are shown on the section on Plate XXI. The idea was also entertained, although not put in practice, of forming similar reheating cores in the discs carrying the moving blades, and these were to be supplied with live steam from a hollow spindle.

About the same time modifications were introduced in connection with the formation of the blades. These had hitherto been cut out of the solid; but independent blades of different brass alloys were introduced in 1891 and adopted in a 100 kilowatt turbogenerator illustrated on Plates XXI. and XXII. The blading is illustrated in Figs. 49 to 54 on Plate XXI. The blades were separately stamped to the desired curvature (Fig. 49), and "upset" at the root (Fig. 50) for the better securing of them in notches in a metal strip (Figs. 52 and 53), which was subsequently calked into a groove formed on the revolving disc and on the plates projecting inwards from the casing on which the guide blades were mounted (Fig. 51). In Fig. 54 there is given a perspective view of the blades thus separately secured in the root strip. The discs and wheels were of cast-iron.

In this turbine[2] made with such blading, in 1891, the dummy piston was also improved by the serration of the sides of the grooves in proximity to which the brass rings on the casing worked. The discs, too, were more firmly secured than in the earlier machines,

[1] Patent No. 5074, A.D. 1891.
[2] Patent No. 10940, A.D. 1891

in which they were held by an end nut. The spindle was constructed with a collar at one end, and into this were screwed long
bolts or studs which passed through holes in the discs and steam seal
or balanced piston, as shown in the section, Fig. 48 on Plate XXI.
A modification was also introduced to increase the sensitiveness of
the action of the solenoid governing gear, a relay, actuated by the
main current from the dynamo, being introduced for the supply of
current to the solenoid, as will be described later, when we review
progress in governing gear.

A further improvement was the construction of a double-sided
disc at the exhaust end of the turbine, so as to secure a greater range
of expansion for the steam before it exhausted into the condenser.
This, as shown in the section, constituted a double-disc outward-flow
turbine placed in the passage communicating with the condenser,
and enabled a large volume of steam to be utilised.

This turbine of 1891, had thus six rotating discs, with blades
on one side, at the high-pressure end, a double-sided disc for low-
pressure steam, steam jackets in the diaphragms for the fixed blades,
adjustment gear for reducing the mean clearance to 0.010 of an inch,
as compared with 0.015 of an inch in the first parallel-flow machine,
serrated grooves for the dummy rings in the balanced piston, and
an improved solenoid governing gear. It was proportioned to drive
a generator to develop 100 kilowatts when running at 4800 revolutions per minute, and created a record for steam consumption, as is
proved by the following excerpts from a report[1] made by Professor
J. A. Ewing, F.R.S., then Professor of Engineering at Cambridge
University, and now the Director of Naval Education at the
Admiralty :—

"The turbine was designed to work with steam at an initial pressure of 140 lb
per square inch, but on the occasion of the trials it was not practicable, for want
of a suitable boiler, to use a pressure of more than 95 lb. per square inch. The first
and second plates (discs), with their turbine blades, were accordingly removed, and
the tests were made with the remaining four small and one large plate. The effect
of this was to make the results of the trial less favourable, as to economy in steam
consumption, than they would have been had the full initial pressure of 140 lb. been
available. In line with the turbine shaft, and directly coupled to it, was the armature

[1] See "ENGINEERING," vol. liii., page 52.

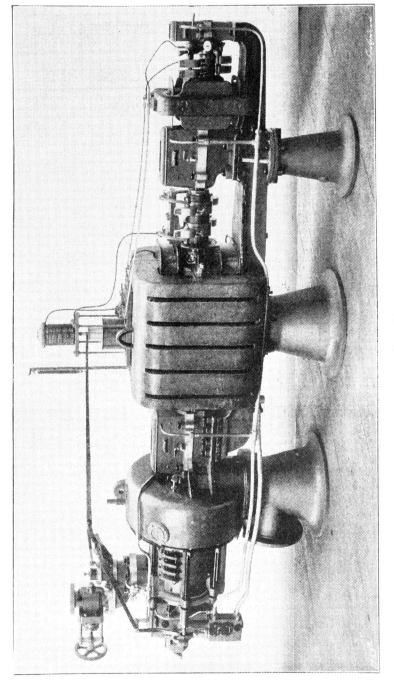

Fig. 55.

Plate XXII.—Perspective View of Outward=Flow Turbine (as on Plate XXI.),
Tested by Professor Ewing.

of an alternate-current dynamo, capable of yielding 100 kilowatts or 100 Board of Trade units of electrical energy per hour, and wound for a potential of 2000 volts. The whole machine, comprising the turbine, dynamo, and exciter, weighed about 4 tons; its length was about 14 ft., and its greatest breadth barely 3 ft. It stood on three cast-iron pedestals, resting on an ordinary concrete floor. There were no holding-down bolts or special foundations, and none seemed to be required. The machine ran almost without vibration.

"The trials extended over three days—December 12th, 14th, and 15th, 1891. The turbine was kept running without change of load, long enough in each case to secure a uniform régime, and to prevent any material error being caused by an inexact reading of the gauge glass on the boiler. The machine ran without any hitch throughout all the trials.

"The consumption of steam by this condensing turbine was 37 lb. per electrical unit generated when the machine was giving its greatest output; the output might be reduced to, say, three-fourths of its greatest value without causing any sensible addition in the consumption of steam per unit; and at half-load the consumption of steam was 39 lb. per unit.

"For the sake of comparison, it may be added that in a good ordinary compound condensing engine of corresponding power the consumption of steam is usually about 20 lb. per indicated horse-power per hour, which (allowing for necessary loss in transmission to the dynamo) corresponds to, say, 36 lb. per unit. The vacuum ranged from $28\frac{1}{2}$ in. at light loads to $26\frac{1}{2}$ in. at full load. The temperature of the cold well varied from 5 deg. Cent. to 18 deg. Cent."

Sir A. B. W. Kennedy, in 1893, also made tests of a turbine of the same design with six high-pressure single discs and one low-pressure disc, the sides of the latter being in parallel. This machine ran at 4800 revolutions, and, driving a continuous-current dynamo producing 110 kilowatts, the steam consumption per kilowatt hour at full load was 27.9 lb.; at half-load, 31.2 lb.; and at quarter-load, 44.1 lb. per kilowatt hour. The steam was superheated to the extent of 66.7 deg. Fahr. in the first test, to 36.6 deg. in the second, and to 20 deg. in the third.[1]

These results, satisfactory as they seemed at the time, were only incentives to further experiment. Careful test showed that although there was a saving in steam consumption by the use of the jackets, it was almost nullified by the volume of live steam used in them. The system of re-heating was therefore abandoned, only one or two machines being completed with this arrangement.

[1] See "ENGINEERING," vol. lvi., page 126.

Experimental work continued. A plain Hero's wheel was tried in 1893, as shown in Fig. 56, on this page. The steam was introduced into a hollow shaft by radial holes on each side. Steam-tightness was secured by the now usual labyrinth packing. The arm was formed of the taper tube of a bicycle fork, and was swaged to a sharp lozenge section to minimise resistance. A divergent jet

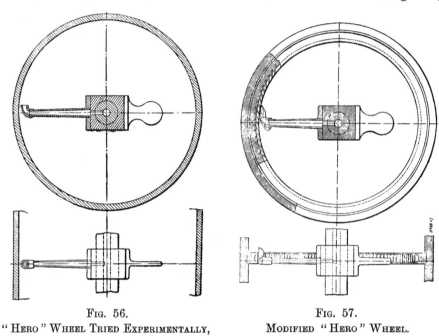

FIG. 56.   FIG. 57.
" HERO " WHEEL TRIED EXPERIMENTALLY,   MODIFIED " HERO " WHEEL.
1893.

was brazed into the outer end, and the larger end was brazed into a steel hub on the hollow shaft. The arm rotated in a vacuum-tight case connected with the condenser. The speed of revolution was 5000 per minute, the distance of the jet from the axis $17\frac{1}{4}$ in., and the lineal speed of the nozzle 780 ft. per second. When driving a dynamo giving 16 kilowatts, the consumption of steam, condensing, was found to be 70 lb. per kilowatt hour—much too high to be satisfactory.

The apparatus was then modified, Fig. 57. A wide brass ring was mounted inside the casing. On the inner surface there was turned a deep elliptical groove, and angular saw cuts were made into which sheet-brass blades were inserted, forming a series of cups,

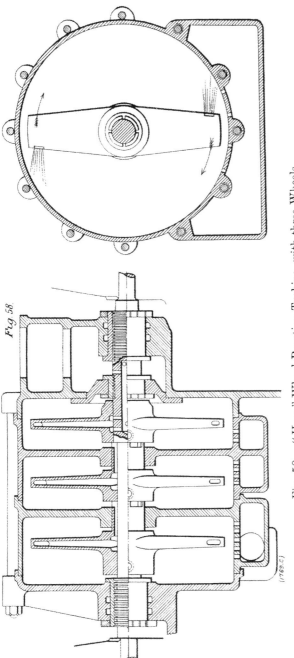

Fig. 58.

(1763.6)

Fig. 58. "Hero" Wheel Reaction Turbine, with three Wheels.

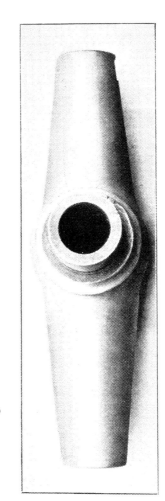

Fig. 59. Perspective View of "Wheel."

Plate XXIII.—"Hero" Wheel Turbine.

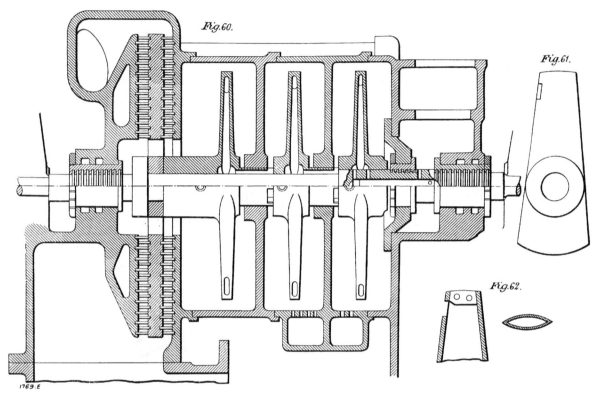

Fig.60.

Fig.61.

Fig.62.

1769.E

FIGS. 60 TO 62. "HERO" WHEEL REACTION AND BLADED DISC TURBINE.

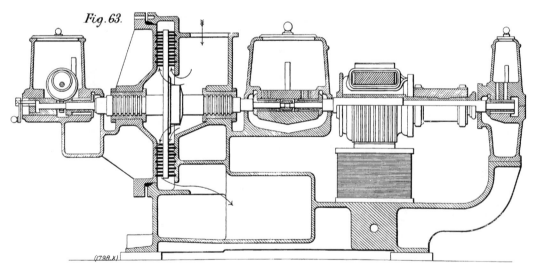

Fig. 63.

(1798.X)

FIG. 63. SINGLE DISC BLADED TURBINE.

Plate XXIV. — Outward = Flow Turbines.

as in the later Reidler-Stumpf turbine. The jet from the end of the arm was arranged to flow tangentially outwards, in order to play into one side of the cups. On the arm were brazed three cupped blades to receive the reacting steam from the stationary cups, and thus it was hoped to attain an efficiency superior to that realised with the "Hero's" wheel with a plain reaction jet.

Difficulties arose owing to the centrifugal forces of the necessarily heavy headpiece carrying the jet and cupped blades, which limited the revolutions. It will be recognised that here again Mr. Parsons came close upon the principle of the system afterwards developed in the parallel-flow turbine of Mr. Curtis; but at this time he was precluded, as already explained, from trying parallel-flow turbines.

Several turbines with double arms and outward-flow jets were made. The first of these "Hero" reaction machines[1] was fitted with three pairs of arms, as shown on Plate XXIII., facing page 42, the perspective view, Fig. 59, showing the spindle which was hollow, with inlets from the steam chest, and outlets into the arms, which were of flattened section to minimise the resistance of the steam within the spindle and in the chamber into which the steam flowed through the nozzles at the ends. For the same reason the outer fish-shaped surface was highly polished. After driving the first pair of arms by reaction the steam re-entered the hollow spindle to flow into the second arms and out of their nozzles, and so on through successive arms to the exhaust port. The nozzles on the arms were arranged with continually increasing area, in order to allow for the expansion of the steam. Drainage apertures and traps were provided for the chambers. Glands, with labyrinth packing, formed steam seals against the inlet and exhaust steam at the ends of the turbine.

Several of these machines were made, some with three and others with five pairs of arms; but these adaptations of the "Hero" wheel were not successful, owing to the frictional resistance of the arms in the chambers.

Tests were also made with a combination of the "Hero" wheel and the bladed disc, as shown on Plate XXIV., on which are

[1] Patent No. 8854, A.D. 1893.

detail sections of arms. The results did not justify the adoption of the system in commercial practice.

Several forms of disc wheels were tried, having a single disc, bladed on one side in some cases and on both sides in others. A section of such a turbine is given in Fig. 63, on Plate XXIV., facing page 43. The leakage losses on these machines were excessive and the steam consumption high, on account of the difficulty of running the sides of a disc in close proximity to a fixed plate projecting from the casing. The test results proved that it was practically impossible to get a single high-speed disc to run within, say, one-hundredth of an inch of a fixed casing owing to the tendency of the disc to oscillate laterally. On the other hand, it was easy to get a spindle with several discs upon it to run within the same distance of its casing with similar lengths between the bearings, there being less chance of lateral vibration or whipping.

It will thus be recognised that during the five years, from 1889 to 1894, when the right to develop the original invention of the parallel-flow turbine was not permitted, valuable work was done, although under somewhat restricted conditions owing to a consciousness that the fundamental basis of research was, at least, of secondary value. The best radial-flow type of turbine was some 12 per cent. less efficient than those of the parallel-flow type of normal economy, working either with or without condensing. For this reason, too, the production of turbines at the Heaton Works was not very extensive, as only outward radial-flow machines could be made. The largest size turbine produced between 1889 and 1894 was of 150 kilowatt capacity, and was for Portsmouth Corporation. The results achieved are recorded in Professor Ewing's report already quoted. It will be noted, however, that as this turbine could not be tested at full designed load owing to reduced steam pressure, the smaller machine tested by Sir A. B. W. Kennedy gave a better economy—the steam consumption being 27.9 lb. per kilowatt-hour.

The research work of the period was, however, invaluable. It established the inefficiency of the " Hero " or " Barker " wheel with single discs, as the efficiency was on an average only about 70 per cent. that of the parallel-flow turbine. It proved, however, that advantages were derivable from the division of turbines into compartments under certain conditions, although this idea was not worked out in a practical form, partly because work on parallel-flow was prohibited. The economy of utilising in low-pressure turbines the exhaust steam from reciprocating engines was recognised, and a patent was taken out in 1894.[1] The scheme has since been applied extensively on land installations, and successfully on several steamships, as will be described in a later chapter. Many details common to all turbines—such, for instance, as the end adjustment of the turbine spindle, blading, dummy balancing pistons, lubricating, governing, and the provision of steam seals—were greatly improved.

The recovery by Mr. Parsons, in 1894, of his patent rights in his original invention was opportune, ánd was of great national importance. All limitations to work on the parallel-flow system were thus at an end, and there were almost immediate developments and improvements. All experiments in connection with radial flow were at once abandoned, as it was recognised throughout the five years' work on this principle that the parallel flow had indubitable advantages. Indeed, this principle dominates all the widely applied turbines of to-day.

One of the first machines of the parallel-flow type made at the Heaton Works was a 350-kilowatt turbo-generator set—a size regarded as enormous at the time. This set, which ran at 3000 revolutions per minute, was made for the Metropolitan Electric Supply Company, Limited, of London. It was a non-condensing machine with parallel-flow rotor within a casing, dummy piston for steam seal, and bearings of the flexible type. The blades in both rotor and casing were not cut from the solid, but were made out of delta metal strips, in which the teeth were cut at the required angle like a comb. These combs, when bent up, gave a complete ring, ready to be put into a groove of the drum or casing, in which they were secured by calking the metal at the side of the ring. Fig. 64, on Plate XXV., facing page 46, shows a part of a ring of blades cut out of the strip for this turbine.

This machine worked at full load, non-condensing, on a steam

[1] Patent No. 367, A.D. 1894.

consumption of 43 lb. per kilowatt hour, which was then regarded as a good result. It was considered, however, that the shape of the blades was unsatisfactory, and a long series of tests was undertaken with various forms of blades, the old turbine "Jumbo" being utilised for the purpose. The idea was to find the best form and curvature for economical working, the most effective stiffening system, and the most secure fastening at the base or root.

It will be remembered that the first blades for parallel-flow turbines were cut on rings as straight paddles at an angle of 45 deg., as shown in Fig. 19, Plate XII., facing page 30. Later these blades were curved at the leading edge, as shown in Fig. 21, on the same Plate. The first blade tried, in comparison with the delta metal strip, Fig. 64, on Plate XXV., facing this page, was stamped out of the solid with a thickened root or expanded foot to enable distance pieces to be dispensed with. This blade is shown in the drawing, Fig. 65, on Plate XXV. The efficiency was fairly satisfactory, but the cost of manufacture was high. Moreover, it was difficult to ensure a good shape at the root in stamping.

Blades and distance pieces calked in a groove were next tried, as illustrated on Figs. 66 to 70, on Plate XXV. Figs. 66 and 67 show the blade and distance piece respectively, the former with its lower edge serrated; Fig. 68 shows a sectional plan and Fig. 69 a section of the blading and wedge pieces in the grooves. On the perspective view (Fig. 70) there is a view of this system of blading as adopted in one of the low-pressure rotors of the turbines made for the Lots Road Station of the Metropolitan District Railway. The distance piece is seen at the end of the groove that is partially filled, ready for the next blade. The longer blades were laced together by a strip fitting in notches cut on the edge of the blades. The ring of blades not completed is without these lacing strips; but the notches are seen, the next row has the lacing strip silver-soldered on. This system showed higher efficiency than the others previously tried, and was applied in all Parsons turbines until the introduction of the "rosary" system, to be referred to presently.

Various sections of such blades were tried, in addition to that shown in Fig. 68. Amongst these was one giving a passage of

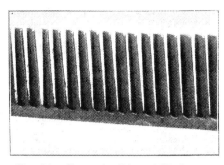

Fig. 64.   Part of a Ring of Blades, cut in Delta Metal Strips.

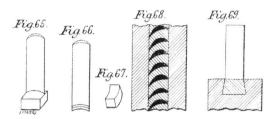

Figs. 65 to 69.   Blades and Distance Pieces tried in 1894.

Fig. 70.   Rotor with Blading and Distance Pieces, as in Figs. 65 to 69.

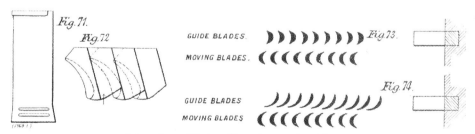

Figs. 71 to 74.   Blades Experimented with in 1900.

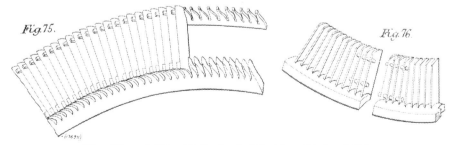

Figs. 75 and 76.   Method of Attaching Blades, 1899.

Plate XXV.—Experiments with Various Systems of Blades (Figs. 64 to 76).

uniform size, as shown in Fig. 73. This gave no advantage, and had the defect that such blades could not be gauged in or out in order to alter the opening or distance between the blades without spoiling the shape of the steam jet.

Blades with the ends bent over in a die were next tried. Fig. 71 on Plate XXV., is an elevation, and Fig. 72 a plan. On the latter the thick lines show the position of the base and the dotted lines the curved blade. Thus the blades overlapped one another, the front edge of one being to the rear of the other adjacent to it. This gave a continuous steam shrouding on the exit side of the ring of blades. The direction of the steam flow is indicated by the arrow. The result showed no distinct advance in economy.

The next tests were made with blades giving passages of uniform size and with the rings of blades in the casing of the ordinary type, curves, and angles, as shown in Fig. 74, and having in some cases a complete shroud at the root and the tips, so as to form effective baffling. The results indicated no improvement.

In the case of some of these tests, notably those in which the blades, as shown in Figs. 73 and 74, were wedged with distance pieces, the method adopted was to form the blades of wood in formers representing casing and rotor on a scale larger than full size. These were placed between glass plates, and water with colouring matter was passed through them. It was at once seen that with a pressure drop across the blades of the impulse type (Fig. 73) the course of the fluid short-circuited, and did not flow tangentially to the surface of the blades. The blade rows in the casing were then removed and replaced by guide, blades with an angle, as in Fig. 74.

The results appeared favourable, and a turbine driving a generator of 75 kilowatts capacity was built in 1895 with this latter blading. In this turbine the guide vanes at the high-pressure end were cast into sector openings occupying a small part of the circumference, and this portion of the turbine worked accordingly with partial admission, a system introduced later in a Continental turbine. At the exhaust end the admission was over the whole circumference. Serrated packing surfaces were provided to minimise leakage. This turbine was at work for several years at the Turbinia

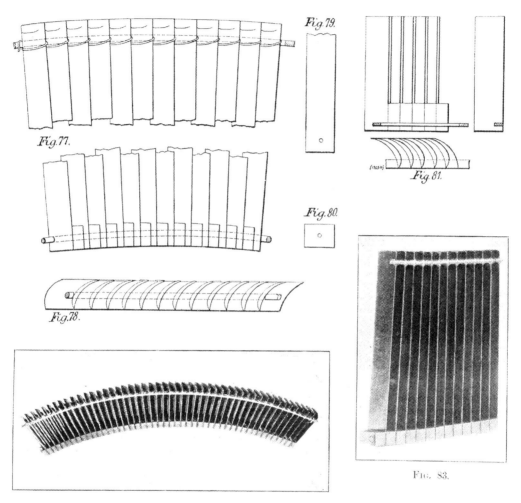

Fig. 77.

Fig. 78.

Fig. 79.

Fig. 80.

Fig. 81.

Fig. 82.

Fig. 83.

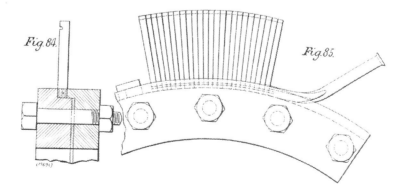

Fig. 84.

Fig. 85.

Plate XXVI.—Latest Systems of Blading. Introduced 1907 (Figs. 77 to 85).

slotted for this purpose; or a side strip, as in Fig. 81, may be used instead of the wire.

The binding is done in segments, of which six or eight form a complete ring, and to facilitate the making of such segment a "former," with a groove, is used, being convex in the case of rotor blading and concave for casing blading, the radius of the curve corresponding, of course, to the rotor or casing for which the blading is intended. Figs. 84 and 85 show such a "former" for rotor blading. When strung alternately on the wire the distance pieces and blades are caulked tightly into the groove, the end of the wire being turned up to secure them. The binding wire is brazed at the ends, the whole of the blading gauged, lacing strips soldered on where necessary, and the segment finished ready for inserting into the grooves in the rotor or casing. This system has the advantage that the building up of the blading in segments can be carried out concurrently with the work of machining the casings and rotors, and this facilitates the progress of construction when the time arrives for the fixing of the blades.

This system has been almost universally adopted in marine work, and to a considerable extent in land work. It is noticeable that this last type of blading is exactly the same as that proposed in 1896, and illustrated in Figs. 66 to 70 on Plate XXV., except that the blades and distance pieces are strung together on a wire in order to hold them together when putting them into the grooves in the casing and rotor instead of each blade and distance piece being put separately in the grooves in casing and rotor.

Since 1905 the ends of the blades have been cut away to a sharp edge like the point of a gouge,[1] as shown in Fig. 70 on Plate XXV., facing page 46, and Fig. 82 on Plate XXVI. This enables the clearance to be reduced to a minimum without danger of excessive friction or damage to the blades in the event of the blades rubbing against the proximate surface. It is a principle in all turbine work that when one moving member is in close proximity to a fixed one, one or other must have a thin edge, otherwise contact results in heating ; consequent inevitable expan-

[1] Patent No. 22127, A.D. 1905.

H

sion of the metal causes still further rubbing, and, as a rule, there is eventually serious damage. Inaccurate adjustment of clearances, after some of the earlier turbines had been put to work, brought about serious injury to the blading—in some cases it was completely stripped. With the sharpened tips now adopted any contact or rubbing causes wear, but no perceptible amount of heat; with wear the contact ceases without damage being done. This thinning away of the tips of the blades is probably one of the greatest improvements in detail made in turbine design in recent years.

This principle is carried out also in the dummies and in the glands; the dummy strip is made with as thin an edge as possible, so that if contact takes place no damage will occur.

A considerable amount of experimental work has also been done in order to determine the most suitable metal for blades in view of the very high temperatures now attained, even with saturated steam. In the early machines the blades were cut out of solid metal—an alloy of copper and tin with only a trace of zinc. The rings were cast and were found unreliable owing to flaws inevitable in the process of casting. As already indicated, delta metal was next utilised, and proved much more reliable. In the radial-flow machines the separate blades used were of ordinary brass, but when the original parallel-flow system was reverted to, delta metal was again adopted, the blades being cut out of the solid, as already described. In 1896, on the introduction of drawn blades, brass was again adopted, and this metal is still used, except for high-pressure turbine blades, where pure copper is preferred, as it is found that the brass is liable to become brittle at temperatures about 460 deg. Fahr., due to traces of impurities in the alloy. Where the steam is highly superheated, nickel bronze, consisting of 20 per cent. of nickel and 80 per cent. of copper, was adopted for a short time (about 1901), but the metal then obtainable proved absolutely unreliable because of the tendency to brittleness; in fact, after a short time the blades were in this respect more like glass than anything else. Steel blades of 25 per cent. of nickel were next tried, but these also were discarded because of their liability to become brittle. The cause of this is not definitely known. In present-day turbines pure copper is almost universally used at the high-pressure end of the turbine, where high temperatures obtain ; but where the steam used is not superheated, and for low-pressure turbines, pure brass is adopted.

Another auxiliary, to which a large amount of attention was devoted, was the governing of the turbine in order to obtain constant speed of voltage, and also to avoid excessive speed. This is a less important element than in the case of a reciprocating engine, because there is less likelihood of serious results following upon the racing of a turbine. The system adopted in the first turbine made has already been described on page 25. It worked very satisfactorily, but, as the governor patents were temporarily lost at the same time as those relating to the turbine proper, new methods had to be devised for regulating the admission of steam in the radial-flow turbines manufactured after 1889. These have been briefly referred to in describing the successive turbines. It will, however, be interesting to have in consecutive order a review of the various developments in governing mechanism.

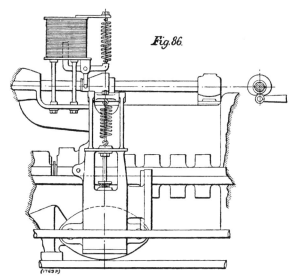

Fig. 86.

FIG. 86. FIRST "GUST" GOVERNOR OF THE TURBINE "JUMBO," 1890.

In 1890 Mr. Parsons patented the method of gust governing,[1] illustrated in Fig. 86, annexed. A sliding sleeve, the position of which was controlled by the governor, was mounted on a second-motion shaft driven by a suitable gearing from the turbine spindle. The outer surface of this sleeve was cut away so as to leave a cam surface, which was used to open or close the circuit through an electric relay which operated the governor valve direct. With the circuit through the relay closed, the governor valve was held open, and

[1] Patent No. 1120, A.D. 1890.

on breaking the circuit the valve closed automatically. The cam surface was graduated in such a fashion, from one end to the other of the sliding sleeve, that if its follower were at one end of the sleeve the circuit remained closed throughout the complete revolution of the cam shaft; whilst if, on the other hand, the opposite end of the cam acted on the follower, the circuit remained broken throughout the revolution, and the valve continued closed. At all intermediate relative positions of the cam and its follower the circuit remained open for a part of the revolution, and was closed for the remainder. The relative period of open and closed circuit varied with the relative positions of the cam and its followers, these being controlled by a solenoid coupled up with the generator driven by the turbine. Thus, as the load on the generator increased the cam was moved, so that the governor valve was held open for a longer proportion of each revolution of the cam shaft. The steam was thus supplied to the turbine in a series of gusts, the duration of each of which increased as the load in the turbine rose.

This governor was seen to be too feeble, and was liable to stick, and modifications were introduced at an early date, a relay operated by fluid pressure being introduced for opening and closing the governor valve.[1] The governors in the first turbines, as described on page 25, *ante*, were of the relay type, using air as a medium in the relay. The pressure of the air was found to be so small that they also were feeble, and liable to seize ; but it was proved a sound principle to have some form of relay between the speed or voltage regulating medium (the centrifugal governor or solenoid) and the steam valve.

A great improvement was introduced when in 1890 there was applied a form of jigging action as provided by the cam in the gear on the " Jumbo " turbine just described, in association with the relay piston, as the jigging action kept everything in motion, and prevented sticking. This feature has characterised the governors since applied to Parsons turbines for land work.

The governor of " Jumbo " with this combination is illustrated in Fig. 87, opposite. It is of the double-beat equilibrium type, having its spindle connected to a spring-loaded piston working in the

[1] Patent No. 11083, A.D. 1890.

relay cylinder. By admitting fluid under pressure beneath this piston, the valve was opened. A leak-off was fitted, so that on the supply of pressure fluid being cut off the valve closed automatically. The supply of pressure fluid to the relay cylinder was controlled electrically. In this arrangement air was still used to operate the relay cylinder. This air was supplied by a reciprocating pump driven from the turbine shaft by friction gearing. A section through the cylinder of this pump is represented at A. The suction

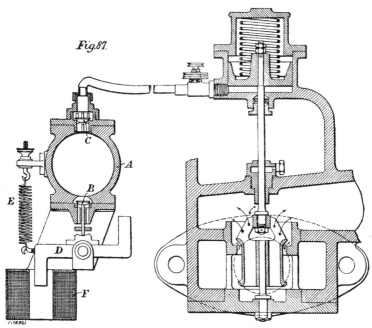

FIG. 87. GOVERNOR OF TURBINES, 1890.

valve to the pump is shown at B, and the delivery valve leading to the relay cylinder at C. If, during the compression stroke of the pump, the valve B were held open no air was delivered into the relay cylinder and the governor valve remained on its seat. On the other hand, with the rocking bar D in the position indicated, the valve B could close, and in that event the pump forced air into the relay cylinder and opened the governor valve. The position of the rocking bar D was controlled by the opposing forces due to the spring E and the attraction of the solenoid F. If, owing to a change in the demand on the generator, the current through the solenoid increased, its attractive force was augmented, and the bar D moved into such a position as to prevent the valve B from completely closing. Consequently less air was delivered into the relay cylinder and less had to leak away before the governor valve could close. Hence the latter closed sooner, and the gust admitted to the turbine was of shorter duration. On the other hand, if by some accident the current completely failed, the spring E came into operation as an emergency governor. It pulled the lever D over to the limit of its travel, in which position, it will be seen, the valve B was held fully open, and no air therefore entered the relay cylinder; the governor valve accordingly remained closed, and all possibility of a dangerous runaway of the turbine was eliminated.

Air relays, however, did not seem satisfactory, and an electrical arrangement was proposed in the following year (1891).[1] In this case the governor merely controlled the closing of the current through the solenoid operating the steam valve, the break being affected by a second pair of contacts, to which therefore all the sparking, and consequent wear, were confined. The arrangement is illustrated in Figs. 88 and 89, Plate XXVII., facing this page.

A long lever having a forked end, fitting under a collar on the stem of the steam valve, carried at its other end a core A of a solenoid. When the current was passed through this solenoid the core was sucked in and the steam valve opened. As it opened, the contact breaker in the opposite end of the long lever struck the flat spring C, separated the two platinum contacts, and thus broke the circuit in the solenoid, and the lever accordingly moved up under the influence of the spring. Contact between the platinum points was thus re-established; but unless another contact, controlled by the governor, was also closed, the current through the solenoid would remain broken. The time at which this secondary contact was closed depended upon the voltage in the generator circuit. If this were low, the governor contacts were closed early, and the core A was then sucked again into its solenoid so quickly that it caught the steam valve before it had actually closed. If, on the other hand, the generator were under-

[1] Patent No. 10940, A.D. 1891.

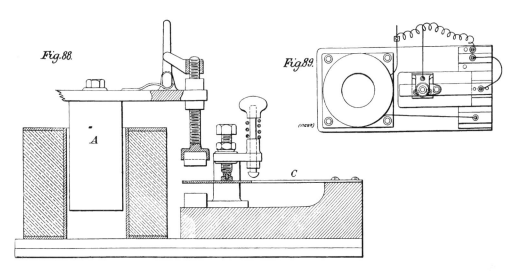

Figs. 88 and 89. Governor of Turbines, 1891.

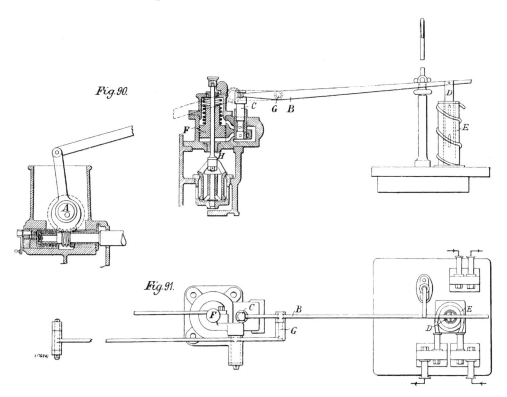

Figs. 90 and 91. Governor of Turbines, 1892 (still in use).

Plate XXVII.—Governors.

loaded, the secondary contact did not re-establish the current through the solenoid until after steam had been completely cut off, and the supply of steam to the turbine was thus accurately adjusted to the load on the dynamo.

This worked well from a governing point of view, but the wear on the platinum contacts was excessive, and, owing to sparking, it was only partly got rid of by a partial series winding on the solenoid. The rate of steam admission, too, was not sufficiently rapid to prevent uneven turning moment and consequent fluctuations in the electric lights run by the turbo-generator. It is probable that the sparking could have been prevented by the introduction of a condenser. It was largely reduced in certain cases by the use of a non-inductive lamp resistance. The difficulty of uneven turning moment, however, prevented this type of governor from being generally adopted. In consequence of this, the relay type was again reverted to ; but, instead of air at a few pounds pressure being used as the actuating medium, steam at boiler pressure was utilised, and in consequence a most satisfactory form of governing[1] was contrived, and is still in use.

This, the latest system, is illustrated in Figs. 90 and 91 on Plate XXVII. Here the eccentric A, driven by worm gearing from the turbine spindle, gives an oscillatory motion to the fulcrum of the floating lever B. At its short end this floating lever carries a small piston valve C, which is set in oscillation, owing to the motion communicated by the eccentric. The position, however, about which it oscillates in its casing, depends upon the position of the core D in the solenoid E. If the current increases through the solenoid, the core is pulled down, raising the other end of the floating lever B ; the mid-position about which the valve C oscillates is therefore raised. This valve C controls the exhaust from the underside of the spring-loaded relay piston F, the steam supply to which passes in by leakage through the bush H. If the mid-point about which the valve C oscillates is- raised, the exhaust from the underside of the relay piston is effected sooner, and in consequence the governor valve closes earlier. If, on the other hand, the governor lowers

[1] Patent No. 15677, A.D. 1892.

the point about which C oscillates, this exhaust from the relay cylinder opens later, and the governor valve admits steam for a longer proportion of the cycle of oscillation. Experience showed, however, the desirability of substituting for the solenoid an ordinary centrifugal governor, which, with the steam relay, has been more used on steam turbines than any other type. The advantage of the governor gear being kept in oscillation is very great, as it prevents all liability of the gear to stick or to hunt. In some types oil under pressure, instead of steam, is used, especially on the Continent.

Thus the governors, up to about the year 1900, were of the electrical type, with the exception of two centrifugal governors fitted to two 100-kilowatt turbines. In some cases compound winding was adopted to enable higher voltage to be obtained at full load than at no load, and in the case of alternators the shunt winding of the governor was across the exciter terminals, while a series winding, carrying the alternating current, with a laminated core, was fitted above to give the necessary compounding. It was, however, found, at an early stage, that it was practically impossible in general station working to run alternators in parallel with electrical governors, and, as a result, mechanical governors have become universal, and although no such difficulties have been met with in the case of continuous-current turbo-generators, the mechanical governor has completely ousted the electrical system.

The first machines to be fitted with mechanical governors were two 100-kilowatt sets for Woolwich in 1896. The Elberfeld machines, to be referred to later, were provided with both mechanical and electrical governors, but after being put to work at Elberfeld the electrical governors were completely detached. For the last six or seven years, the battle of electrical *versus* mechanical governing has been completely decided in favour of the mechanical system.

The governing arrangement of the present day is identical with that shown in Figs. 90 and 91, on Plate XXVII., facing page 54, with the exception that the floating fulcrum G has been dispensed with and the jigging motion is given by an oblique cam on the governor sleeve, as mentioned in the case of an earlier governor (page 51), whilst an ordinary mechanical governor is substituted for the solenoid. The relay is, to all intents and purposes, excepting in matters of detail, as shown in Fig. 90.

With regard to the effect of intermittent admission on the consumption of steam, there is little advantage even with superheated steam, unless the gusts are very slow with a long interval after each. In fact, with saturated steam no effect was appreciable between puff governing and throttle governing. As parallel running of alternators has necessitated more and more even turning moment the rapidity of the gusts has had to be increased. When first the modern type of governor was put in hand the frequency of gusts was about 160 to 180 per minute. This has, however, been increased to from 350 to 400, in order to give more even turning moment for parallel running. In many cases bad parallel running has been cured by increasing the number of gusts from, say, about 180 or 200 to some 400. The reason why gust governing does not give better economy is, of course, the condensation which takes place in the cylinder at the beginning of each puff, if the steam is not superheated ; even when it is superheated loss is caused by the transfer of heat alternately from the steam to the casing and rotor and back again, without doing work. The surface of the turbines and the blades is so enormous that heat is very quickly transferred from the metal to the steam, and *vice versâ*. The reason why puff governing has been found preferable to throttle governing is that all the parts of the governor, as already mentioned, are kept free, so that no sticking or hunting can take place.

Another subject which has had continuous attention is that of lubrication. The method adopted with the first turbine has been fully explained on page 24. Subsequent to this, an oil pump, generally of the reciprocating type, was usually adopted, but in a few machines there was no oil pump at all, the bearings being fed from small tanks on the bearings filled by hand. This, however, did not prove satisfactory. In more recent machines the oil is circulated through an oil tank by a pump, usually of the rotary type, although sometimes of the ordinary reciprocating type, driven by worm wheel from the turbine shaft. After passing through a cooler, the oil is forced

I

through the various bearings at pressures ranging from 2 lb. to 7 lb. per square inch, according to circumstances, and thus there is ensured a perfectly automatic system of lubrication.

In 1903 Mr. Parsons first applied his vacuum augmenter[1] to condensers, and this is one of the important improvements in turbine installations in recent years. It had been recognised from the beginning of turbine work that high vacua in condensers offered considerable advantage, as the turbine could use steam expanded beyond the limits practicable in a reciprocating engine. The ratio of cylinders and proportions of exhaust ports and eduction pipes usual in the latter give an exhaust pressure of from 7 lb. to 16 lb. absolute, so that there is little to be gained in such engines by using a vacuum higher than about 26 in. On the other hand, in the steam turbine vacua of $28\frac{1}{2}$ in. or 29 in., equal to absolute pressures of from $\frac{3}{4}$ lb. to $\frac{1}{2}$ lb., can be utilised as expansion continues to a pressure corre-

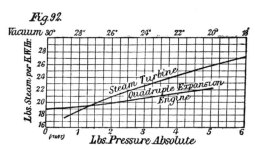

FIG. 92. DIAGRAM SHOWING EFFECT OF HIGH VACUA ON STEAM CONSUMPTION.

sponding to the highest vacuum attainable, since the difficulties due to large volumes of steam do not occur as in the case of piston engines. An advance from 26 in. to 28 in. vacuum reduces the steam consumption by only 2 per cent. in the case of a piston engine as compared with 10 per cent. in the turbine.

The increase in economy due to high vacua is well shown in the diagram (Fig. 92), which was prepared by Professor Weighton, of the Armstrong College, Newcastle-on-Tyne, as the result of tests of quadruple expansion condensing engines there.[2] One curve shows the steam consumption in pounds per kilowatt hour for a steam turbine, and another curve the same result for the piston engine, deduced from the brake horse-power, and assuming an overall efficiency of dynamo of 93 per cent.

[1] Patent No. 840, A.D. 1902.
[2] "Transactions of the Institute of Naval Architects," vol. l., page 94.

As early as 1893 Mr. Parsons gave special attention to the designs of condensers, aiming generally at the exhaustion of air from the condenser by the use of separate pumps of large capacity. In 1896 there was adopted a weir at the bottom of the condenser to retain sufficient of the condensed water to submerge three or four rows of tubes, counting from the bottom upwards.

While the results were good, it was recognised that still higher efficiency in condensers was attainable. Careful collation of the results and their analysis established that the difference between the

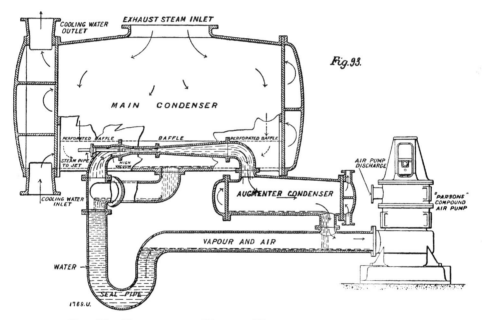

FIG. 93. DIAGRAMMATIC VIEW OF VACUUM AUGMENTER SYSTEM.

temperature of the outlet water from the condenser and that due to the vacuum should not exceed 5 deg. Fahr. when condensing at the rate of 12 lb. of steam per hour per square foot of cooling surface. A principal cause of excess is the resistance offered to heat passing from the steam to the cooling water, and this is most largely affected by the quantity of air in the condenser. Experience has shown that the diameter and disposition of tubes, the flow of steam, and the amount and the velocity and "turbulence"—in contradistinction to quiescent flow—of the circulating water, are important elements in the attainment of efficiency, since they affect the deposit of foreign matter on the tubes from the circulating water. But air, if present, prevents the steam from coming into contact with the cooling surface, and thus, in large measure, nullifies other improvements, and proves perhaps the worst obstacle to the attainment of high efficiency. To diminish the residual air in the condenser the vacuum augmenter was introduced; it is practically a dry-air pump of very large capacity and high efficiency, with the minimum of steam consumption.

It will be seen from the diagrammatic section, Fig. 93, that the air and vapour is compressed by a steam jet in the contracted pipe to about half of its bulk, and is delivered to the air pump at a pressure about 2 in. higher than the pressure in the condenser through an auxiliary condenser having from 2 to 3 per cent. of the cooling surface of the main condenser. In this auxiliary condenser the air is cooled, and the vapour partially condenses before it enters the pump. By the extraction of practically all the air the conductivity of the cooling surface is much increased.

The steam used by the vacuum augmenter jet is about 0.6 per cent. of the steam used by the turbine plant at its normal full load. In Table V. there are given figures obtained from four typical condensers in commercial use, having cooling surfaces ranging from 750 to 7700 square feet, both with and without the vacuum augmenter in use.

In the case of condenser " B " a high degree of air tightness is shown, with a condensation rate of between $12\frac{1}{2}$ lb. and 13 lb. per hour per square foot of cooling surface, and about 900 lb. of cooling water per hour per square foot of surface. The difference $T - t_1$ between the temperature due to the vacuum obtained and the cooling water supply being only 20 deg. Fahr. with the augmenter in use, and 26.9 deg. Fahr. without, and the gain in vacuum of 0.19 in. over that without the augmenter in use.

In the condenser " C," with a condensation rate of $4\frac{3}{4}$ lb. per hour per square foot of surface there is a much less degree of air tightness, $T - t_1$ being 40 deg. Fahr. without, and 13.9 deg. Fahr. with, the augmenter jet in use, and a corresponding gain of vacuum of 1 in. of mercury. This shows the great advantage of such a vacuum augmenter where air leaks are present.

TABLE V.—TEMPERATURES AND PARTICULARS OF CONDENSERS WITH VACUUM AUGMENTER.

Condenser.	A		B		C		D	
—	Augmenter in Use.	Augmenter Off.	Augmenter in Use.	Augmenter Off.	Augmenter in Use.	Augmenter Off.	Augmenter in Use.	Augmenter Off.
Cooling water:—								
Inlet t_1 (deg. F.)...	41	41	50.5	50.5	56.1	57	77	77
Outlet t_2 (deg. F.)	55	55	64.3	65.2	63	64.5	93	93
$t_2 - t_1$ (deg. F.) ...	14	14	13.8	14.7	6.9	7.5	16	16
Cooling water per square foot of surface (pounds per hour)...	696	680	935	855	686	631	522	522
Condenser:—								
Measured temperature of steam space (deg. F.)	73.5	80	74.3	80.4	71	102	—	—
Measured vacuum at 30 in. bar. (in.)	29.26	28.97	29.25	29.06	29.25	28.25	27.9	27.5
Temperature due to vacuum measured (deg. F.) (T.)	69.5	80.3	70.5	77.4	70	97	102.8	109
$T - t_1$ (deg. F.) ...	28.5	39.3	20	26.9	13.9	40	25.8	32
Air pump discharge:— Temperature (deg. F.) ...	66	59	67.5	64.75	70	76	98	—
Total steam (pounds per hour) ...	7300	7100	28,500	27,750	3370	3360	64,100	64,100
Steam per square foot (pounds per hour)	9.75	9.5	12.9	12.57	4.75	4.74	8.35	8.35
$\dfrac{t_2 - t_1}{T - t_1}$ ...	.491	.356	.69	.545	.496	.187	.62	.5

Accepting $\dfrac{t_2 - t_1}{T - t_1}$ as the formula for the thermal efficiency of a condenser, where t_1 and t_2 are the temperature of the inlet and outlet water respectively, and T the temperature due to vacuum, it may be said that large condensers in electric light stations have shown efficiencies as high as 90 per cent., and in marine work as 85 per cent.

Another view is that the efficiency of the condenser should be measured according to the average British Thermal units transmitted per square foot of cooling surface per hour per 1 deg. Fahr. difference of temperature. In ordinary condensers this unit does not exceed from 250 to 300, and is often lower, but with the augmenter it may be increased to between 800 and 1000. Another way of determining efficiency is to proceed on the lines that the maximum vacuum which can be obtained in the condenser is that due to the outlet water. The temperature of the inlet water is generally fixed by circumstances the engineer cannot control, and it is not convenient in practice for the volume of circulating water to be more than sixty times that of the condensed steam, which means a rise of some 17 deg. Fahr. in the circulating water. It is thus desirable to have a temperature due to the vacuum as close to this outlet temperature as possible, and in modern condensers, fitted with augmenters, this difference can be kept as low as 5 deg. to 6 deg. Fahr. when condensing, say, about 12 lb. of steam per square foot of cooling surface per hour, while in ordinary condensers the temperature difference often exceeds 20 deg. Fahr. Since every 3 deg. Fahr. represents approximately 1 per cent. in the steam consumption of the turbine, the importance of keeping the loss as low as possible will be clearly appreciated. In this part of the work Mr. Gerald Stoney, Mr. S. S. Cook, and Mr. R. Howe took a large share.

During the progress of the experimental work on such details of design as blading, governing, lubricating, and the application of a satisfactory condenser to secure high vacua, manufacturing was, of course, actively continued, and progress was rapid from 1894, when the proprietary rights in the parallel-flow turbine patents were recovered. Up to that time the largest turbine made was built for driving a 150-kilowatt generator. During the five succeeding years (1894-99) the total output of turbines, excluding marine work, was equal to about 10,000-kilowatt capacity, as compared with 4200 kilowatts during the five years, 1884-89, when Mr. Parsons was engaged at Gateshead in the production of the parallel-flow turbine. In the period between 1889 and 1894 the output of turbines was comparatively small, due to the fact that Mr. Parsons was occupied at his works at Heaton in the attempt to produce a satisfactory radial-flow turbine.

The result of the improvements effected in turbine design, following upon the recovery of the patent rights, are reflected by a comparison of the efficiency established by two tests, one in 1896 and the other in 1900.

The first results were arrived at by Mr. William D. Hunter, M.I.E.E., the Engineer to the Newcastle and District Electric

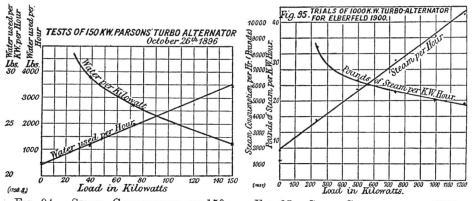

FIG. 94. STEAM CONSUMPTION OF 150-KILOWATT TURBO-ALTERNATOR, 1896.   FIG. 95. STEAM CONSUMPTION OF 1000-KILOWATT TURBO-ALTERNATOR, 1900.

Lighting Company. These tests were made with the compound steam turbine, making 9400 revolutions per minute, and geared in the ratio of 2 : 1 by helical-spur gearing to a 150-kilowatt alternator and direct-coupled exciters. The steam pressure in the turbine receiver was 70 lb. per square inch not superheated, the vacuum $26\frac{1}{8}$ in., with a 29.3-in. barometer, and the water consumption at full load $22\frac{1}{4}$ lb. per kilowatt hour.[1] The diagram, Fig. 94, gives the water consumption per hour and per kilowatt hour at various loads.

The two Elberfeld machines, of 1000-kilowatt capacity, made in 1900, greatly excelled even this highly economical result. Indeed, the consumption of steam — 19 lb. per kilowatt hour — was then unexcelled by any engine, and the diagrams of results on Fig. 95

[1] See "ENGINEERING," vol. lxii., page 625.

form a suggestive contrast with the performance of 1896, recorded in a similar curve. The Elberfeld machines corresponded to the design made in 1889 (Plates XIV. and XV.), but at that time the low-pressure turbine was not manufactured. The results were so remarkable—they are not much excelled even to-day—that the sectional drawings reproduced on Plate XXVII. will be studied with interest. On Plate XXVIII. there is a perspective view. The high-pressure and low-pressure turbines were mounted tandem-wise in one line along with the electric generator.

The City Corporation of Elberfeld, in Germany, for whose Electrical Supply Works the turbines were constructed, laid down stringent conditions and appointed a commission to determine by tests whether these were complied with. Mr. W. H. Lindley, M. Inst. C.E., one of the members, deserves great credit for his enterprise in recommending the Corporation to take the bold step of ordering such a powerful compound turbine when other engineers hesitated because of the lack of practical experience with this development of the system. The other members of the commission were Herr M. Schröter, Professor of Mechanical Engineering at the Polytechnicum, Munich, and Dr. H. F. Weber, Professor of Physics at the Polytechnicum, Zurich.

From the report they made[1] we take one or two extracts, as they establish (1) the faith of the inventor in his work as reflected in his guarantees, and (2) the practical success achieved alike in design and manufacture by the reliability, regularity, and economy of working proved by the tests.

The specification required that each machine should have an output of 1000 kilowatts useful effect at 4000 volts, and with fifty complete cycles per second measured on an inductive load with $\cos \phi = .8$. The speed of the turbine was to be 1500 revolutions per minute, and the admission of steam to the steam chest once in each eight revolutions of the turbine, or 187.5 admissions per minute, so as to synchronise with the revolutions of a reciprocating engine in use at the power station. The speed variation when the load was increased gradually from no load to full load was not to

See "ENGINEERING," vol. lxx., page 830.

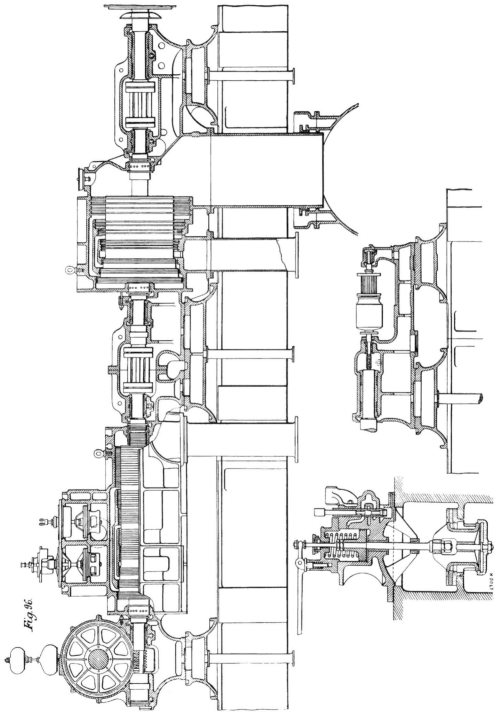

Fig. 96.

Plate XXVIII.— 1000=Kilowatt Turbo=Alternator for Elberfeld, 1900 (Fig. 96).

Plate XXIX.—1000=Kilowatt Turbo=Alternator for Elberfeld, 1900 (Fig. 97).

exceed 4 per cent. ; with a variation of 25 per cent. of the actual load suddenly thrown on or off the speed variation was not to exceed 0.8 per cent., provided the total load was not less than 200 kilowatts. An electric governor was to keep the voltage within 1 per cent., with a variation of 25 per cent. of the actual load suddenly thrown on or off. With at least eleven atmospheres absolute pressure and 50 deg. Cent. superheat of steam at the stop valve, and with circulating water of not more than 18 deg. Cent., supplied at the rate of 430 cubic metres per hour for full load, the steam consumption per net kilowatt hour, measured by the alternating current at the switchboard, was not to exceed 11 kilogrammes ($24\frac{1}{4}$ lb.) at full load, 11.3 kilogrammes (24.9 lb.) at three-quarter load, 12 kilogrammes (26.45 lb.) at half load, and 14 kilogrammes (30.8 lb.) at quarter load. The consumption of steam at no load was not to exceed 1060 kilogrammes (2337 lb.) per hour. The actual steam consumption on the official tests are given in Table VI., on the next page.

It will be seen that the guarantees were very considerably exceeded, except at quarter-load. At full load they were 1.81 kilogrammes (4 lb.), at three-quarter load 1.31 kilogrammes (2.9 lb.), below the stipulated amount when the figures are corrected for the lower degree of superheat at which the trials were run, as shown in the second part of the Table on the next page.

The average variation of speed when the load was gradually altered from nothing to full, and *vice versâ*, was 3.6 per cent. When a part of the load varying from 16 to 63 per cent. was suddenly switched off, and the regulation was entrusted to the centrifugal governor, the revolutions varied from 1 to 1.9 per cent. immediately after the sudden change of load, whilst the permanent variation in speed amounted to from 0.4 to 1.3 per cent. The electrical governor under similar conditions kept the average variation of potential within 1.1 per cent. of the initial potential. Between no load and full load the drop of potential on a non-inductive load was only 1.02 per cent., or 20 per cent. of that permitted by the speed. The drop of potential with inductive load amounted to 11 per cent. between no load and 1000 kilowatts.

K

TABLE VI.— STEAM TRIALS OF ELBERFELD COMPOUND TURBINES AND
ALTERNATORS, 1900.

—	Exact Value of Output.	Steam Consumption per Kilowatt Hour.		Steam Consumption in One Hour.
	kilowatts	lb.	kilogrammes	kilogrammes
Preliminary trial	1172.7	18.22	8.26	9,689
Overload...	1190.1	19.43	8.81	10,485
Normal load	994.8	20.15	9.14	9,092
Three-quarter load	745.3	22.31	10.12	7,542
Half load	498.7	25.20	11.42	5,695
Quarter load	240.5	33.76	15.31	3,774
No load, with alternator excited ...	0	...	...	1,844
No load, without excitation	0	...	...	1,183

The tests were made with an average superheat of 14.3 deg. Cent. (25.7 deg. Fahr.) instead of 50 deg. Cent. (90 deg. Fahr.) as stipulated for. Better figures might have been obtained with a higher degree of superheat. When the figures are corrected to the average superheat of 14.3 deg. Cent. they stand as follows :—

Load.	Guarantee at 50 deg. Superheat.	Consumption at 14.3 deg. Superheat.		Difference.
kilowatts	kilogrammes per kilowatt hour.	kilogrammes per kilowatt hour.	lb.	kilogrammes per kilowatt hour.
1250	...	8.63	19.0	...
1000	11.0	9.19	20.2	− 1.81
750	11.3	9.99	22.0	− 1.31
500	12.0	11.41	25.1	− 0.59
250	14.0	15.28	33.7	+ 1.28

Thus on thoroughly exhaustive trials by an international board the turbine worked with a water consumption per hour of 19 lb. per kilowatt hour, delivered at the switchboard—a result which had not then been attained by any reciprocating plant—and in all other respects proved equal to the very severe conditions required by the contract.

Since 1900, increase in size and improvement in manufacture and in details of design, as already reviewed in connection with experiments on blading, governors, lubricating, condensing plant, &c., have resulted in a reduction of the steam consumption of turbo-generators to 13.2 lb. per kilowatt hour, with a boiler pressure of 200 lb., a vacuum of 28.8 in., and 120 deg. Fahr. of superheat.

The development since the Elberfeld machine so clearly demonstrated the efficiency of the system has been more a matter of detail, and also of increasing size and power. The tandem arrangement, to which the Elberfeld machine belonged, was soon abandoned, as it was thought to be very expensive in manufacture. As soon as it became possible to obtain satisfactory steel castings of large size for turbine rotor drums, it was easy to have the whole of the expansion in a single casing enclosing the high and low-pressure turbines, capable of running a 4000-kilowatt generator. A single-casing machine of this size was made for the Carville Station of the Newcastle Electric Supply Company in 1903, and later a turbo-alternator of 5000-kilowatt capacity for Sydney, in Australia.

Recently, however, there has been a tendency to revert to the Elberfeld design, having separate casings for high-pressure and low-pressure turbines, as is exemplified by the 5000-kilowatt set recently installed at the Lots Road Generating Station of the Underground Railways of London and the 6000-kilowatt set for the Rand. The features of these designs will, however, be dealt with in a subsequent chapter reviewing the development of turbine electric generators and descriptive of the present-day Parsons land turbines and electrical generators.

With the frequent attainment of steam consumptions of 13 lb. to 13½ lb. per kilowatt hour where moderate degrees of superheat are adopted, there is no room for surprise that the turbine has taken first place, not only for electric generation but for many other purposes, and that in ten years the maximum size has increased from 500 to over 6000 kilowatts in land practice, and from 2300 to 70,000 shaft horse-power for ship propulsion; but the story of the evolution of the marine type merits a separate narrative. We may therefore conclude this review of the experimental work upon the mechanical details of turbines with the remark that the student of engineering history will recognise that, in addition to evolving an entirely new prime mover, Mr. Parsons and his colleagues have made a large contribution to our knowledge as to the application of fluid pressures in practical work, while the record of ingenuity exercised in devising methods for converting potential energy into mechanical energy must prove fascinating, if not stimulating, to all engaged in mechanical science.

THE EVOLUTION OF THE MARINE TURBINE AND ITS APPLICATIONS TO BRITISH WARSHIPS.

THE suitability of the steam turbine for the driving of the screw propellers of steamships was recognised as soon as the first turbo-dynamo was found to be a practical success. It was seen, however, that there were serious difficulties to be overcome before success could be achieved in the application of the system. These were principally associated with the fact that the screw propeller, as then designed, was not efficient when run at the high rate of revolution necessary for turbine efficiency, which latter is largely dependent on the highest possible ratio of blade speed to steam velocity.

Mr. Parsons began the study of the marine problem at the Gateshead Works, where the early turbines for dynamo driving were made, and where he had produced the design for a compound turbine of the tandem type, as adopted in the Elberfeld machine, built ten years later, as described in the preceding chapter. His first design of a machine for driving a screw propeller was also of the tandem type, of 200 horse-power, with two separate turbine casings. About half the expansion was to take place in each turbine, and the steam was to be carried at about atmospheric pressure from one to the other by a pipe between the turbines. It was estimated that this engine would consume about 13 lb. of steam per equivalent indicated horse-power when applied to drive a propeller. This engine, however, was never made, because of the partnership difficulties already alluded to.

In 1891 the first condensing turbine—of the radial-flow type—was made and tested at Parsons Works at Heaton, and the steam consumption was found to be about 27 lb. per kilowatt hour, or about 18 lb. per shaft horse-power. This result was an incentive to experimental investigation of propeller problems. Indeed, serious consideration was then given to the proposal to build an experimental turbine-driven vessel to attain the highest speed possible, notwithstanding that it was recognised that the radial-flow turbine was less efficient than the parallel-flow machine which Mr. Parsons was then precluded from manufacturing, as he did not at that time possess the rights of his earliest patents.

Ship-shape models about 2 ft. long were made in 1894, and towed by a fishing-rod and line in a small pond at Ryton-on-Tyne, where Mr. Parsons resided. Models with rounded sterns when run at high speed showed a tendency to sit up. This led to the adoption, in the 2 ft. model, illustrated on the Plate opposite, of the flat stern, introduced earlier by Sir (then Mr.) John I. Thornycroft. The model was fitted with clockwork and a powerful twisted rubber spring, which drove a two-bladed propeller half an inch in diameter and about one pitch ratio. The speed obtained was about 6 knots, with about 18,000 revolutions of the propeller. Of this model two perspective views are given on Plate XXX. (Fig. 98). Although simple in its construction, it yielded suggestive results.

The next model was 6 ft. long, and was driven also by twisted rubber cord, but it had single reduction spiral gearing on the propeller shaft, as shown in the perspective views of this model on Plate XXX. (Fig. 99), and on the section (Fig. 100). The working torque delivered to the propeller was determined by substituting for the propeller a light fan surrounded by a cardboard box with internal vanes. This box was mounted on an almost frictionless axis, coaxial with the screw shaft, and formed an air torsional dynamometer. The working speed of the propeller of this model was 8000 revolutions per minute, and the vanes of the air dynamometer were adjusted so that when it replaced the screw approximately the same number of revolutions were obtained. The torque was then measured by a weighted lever attached to the box. By this means it was possible to eliminate the losses due to friction in the gear shafting and the losses in the intertwisting rubber at the working speed. The resistance of the model was determined by towing it in a pond at the Heaton Works. The towing apparatus consisted of a wheel and axle driven by weights, one with a small

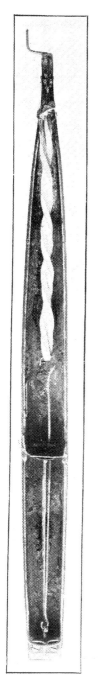

Fig. 98. Plan and Elevation View of First (2 ft.) Model of "Turbinia."

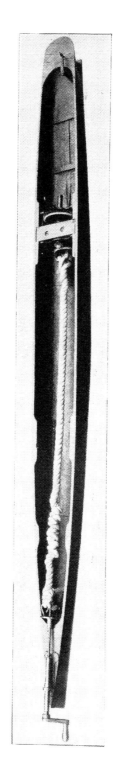

Fig. 99 View of Second (6 ft.) Model of "Turbinia."

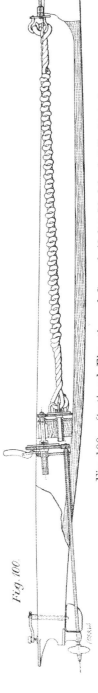

Fig. 100.

Fig. 100. Sectional Elevation of Second (6 ft.) Model of "Turbinia."

Plate XXX.— Experimental Models.

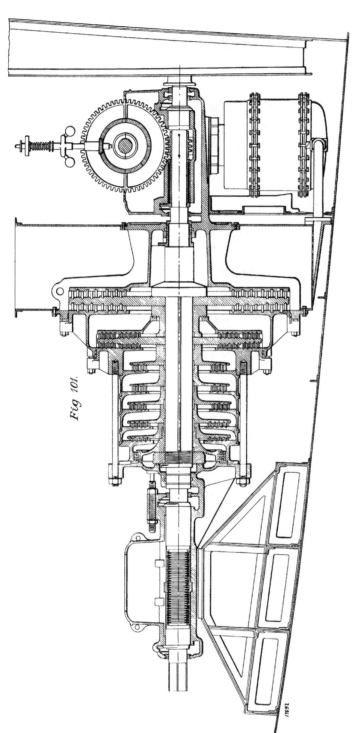

Fig 101.

Fig. 101.   First Radial-Flow Turbine of the "Turbinia."

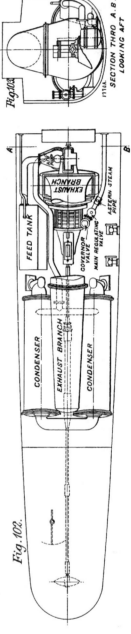

Fig.103.

SECTION THRO A.B.
LOOKING AFT

FEED TANK

EXHAUST
BRANCH

GOVERNOR
VALVE

MAIN REGULATING
VALVE

ASTERN STEAM
PIPE

CONDENSER

EXHAUST BRANCH

CONDENSER

Fig.102.

Figs. 102 and 103.   General Arrangement.

Plate XXXI.—First Turbine Machinery of the "Turbinia."

lift for starting and the other for steadying the drive and thus ensuring a constant speed over the run. The salmon line used around this wheel for towage had two tags, 30 ft. apart, and the model's speed was determined from the time which elapsed between the passing of the first and second tags opposite a guide post. From this the tow rope resistance was deduced.

From these measurements the efficiency of a propeller of normal proportions, running at 8000 revolutions, was calculated. The slip ratio was estimated approximately by close observation of the times between treble and double knotting of the rubber, which was found to represent a definite number of revolutions. All the observations were repeated many times with fairly close agreement.

The result showed a good efficiency of propeller after allowance for such decrease as arose from the augmentation of skin friction resistance due to the small scale of the propeller. Corroboration of the accuracy of this simple and ingenious apparatus was afforded a few years later. The 6-ft. model was developed into the "Turbinia," a vessel of torpedo boat type. The effective horse-power was calculated from the early experiments, and it is worth noting that, when the effective horse-power realised by the "Turbinia" was tested by models, three years later, in the Government tank at Portsmouth, the difference between the two determinations was only from 2 to 3 per cent. This proof of the accuracy of the original experiments is the more creditable in view of the simplicity of the apparatus.

The results, alike as regards turbine design and propeller efficiency, justified the formation of a small syndicate for the building of a boat, which subsequently became famous as the "Turbinia." The vessel was constructed at works established by the syndicate at Wallsend-on-Tyne, whence have since come many of the finest marine turbines made. The dimensions of the hull were 100 ft. length, 9 ft. beam, and 3 ft. draught, at which the displacement was 44 tons. A double-ended boiler was fitted, having 1100 square feet of heating surface and 42 square feet of grate area. It was of the straight-tube type, the tubes being $\frac{1}{2}$ in. internal diameter with $\frac{1}{16}$ in. thickness of metal; two return water tubes of 7 in. diameter

connected the top drum at each end with the water pockets. The
diameter of the top drum was 34 in. The first turbine fitted was
of the radial-flow type, as the proprietary rights in the parallel-
flow machine had not then been recovered; the compound principle
was applied, and the machine drove a single shaft. A section of
this turbine is given in Fig. 101, on Plate XXXI. Its main
features agree with the outward-flow turbines then made for driving
dynamos and other duty, as described in the preceding chapter.
The compound arrangement, as well as the simple form of ball
governor, gland packing, bearings, and thrust rings, are clearly
shown on the section. The arrangement of the machinery in
the "Turbinia" is shown by the plan and section (Figs. 102 and
103) on the same Plate. It will be seen that the air pump was
driven by gearing from the turbine spindle, and that the high-
pressure end was aft, and the low-pressure turbine and the reversing
blades forward, while the eduction or exhaust pipe extended aft over
the turbine to the condensers in the wings on each side of the shaft.

The preliminary trial of the "Turbinia" was made on
November 14, 1894, but before a satisfactory arrangement of propellers was arrived at, seven different designs were tried, the vessel
going to sea on thirty-one full-speed tests. The first—a single—
propeller was a two-bladed screw of 30 in. diameter and 27 in.
pitch, which gave excessive slip—48.8 per cent. when running at
1730 revolutions. A single four-bladed propeller making 1600 revolutions per minute, gave similarly unsatisfactory results. Multiple
propellers were next tried, set on the shaft about three diameters
apart. The most favourable results were obtained with treble
screws, with diameters of 20 in., 22 in., and 22 in. respectively, and
one pitch ratio; these brought the mean slip down to 37.5 per cent.,
the revolutions being 1780 and the speed $19\frac{3}{4}$ knots. But this
result was still disappointing.

The trouble was ultimately diagnosed as cavitation—a term
applied to a phenomenon first noticed by Sir John Thornycroft
and Mr. S. W. Barnaby in 1893, during the trials of the "Daring,'
one of the first of the 27-knot torpedo-boat destroyers. It consists
in the hollowing out of the water and the production of vacuous

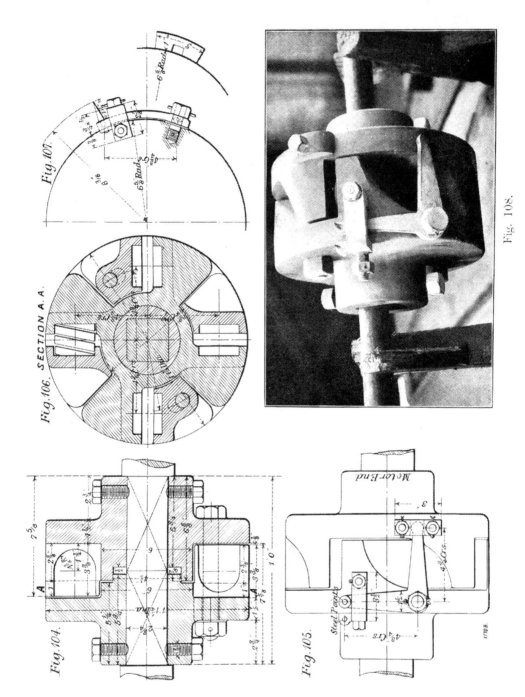

Fig. 107.

Fig. 106. SECTION A.A.

Fig. 108.

Fig. 104.

Fig. 105.

Motor End

Steel Points

Plate XXXII.—Shaft Dynamometer (Figs. 104 to 108).

cavities by the force exerted by the blades revolving through the water at high speed. Cavitation could not exist in the model experiments, and the serious limitations imposed were not realised until many trials had been made with the "Turbinia." Therefore, when the "Turbinia" failed to approach the speed expected, investigation was commenced to discover where the defect lay.

A shaft dynamometer was made; it is illustrated on Plate XXXII. as it is the forerunner of later torsion meters. The ordinary shaft coupling was replaced by two half couplings, the one carrying four lugs and the other having four recesses, in which were strong spiral springs so arranged that the transmitted torque compressed the springs; the relative angular displacement of the couplings measured the compression and therefore the torque on the shaft. A bell-crank lever was pivoted on the edge of one coupling and engaged in a shoe on the other. The other arm of the bell-crank carried a sharp pointer. To the coupling carrying the pivot was fixed another pointer of the same length. When the couplings were rotating a piece of soft wood was held so that their points scratched lines on it, and by the distance between these lines the torque could be calculated. This dynamometer was calibrated by weighted levers before and after use, as is now done in all shaft dynamometers or torsion meters. The maximum shaft horse-power thus registered was 960 at 2400 revolutions, and as the effective horse-power required to drive the "Turbinia" was known from the model experiments previously described, it was made obvious that the propellers were very inefficient.

It was then decided to replace the single shaft by three shafts driven by three turbines in series, and to make the revolutions 2200 per minute, while maintaining the same coefficient of efficiency in the turbines. By this time—towards the end of 1895—the proprietary rights in the parallel-flow turbine had been recovered, and the new turbines were therefore made of the more efficient type. Part-elevations and part-sections of the high-pressure and inter-mediate-pressure turbines are given on Plate XXXIII. (Figs. 109 and 110), and of the low pressure ahead, and of the astern turbines in Plate XXXIV. (Figs. 111 and 112). On Plate XXXIV. there

L

is also shown the arrangement and cross-section of the machinery
in the ship (Figs. 113 and 114). The high-pressure turbine was
on the starboard shaft, the intermediate on the port shaft, and
the low-pressure on the centre shaft, the steam doing work
successively in each turbine. The separate turbine for driving
the ship astern was on the centre shaft. The drawings are self-
explanatory in the light of the descriptive details of land turbines
in the preceding chapter. These marine turbines were made shorter
and of considerably greater diameter than land turbines in order
to reduce the revolutions. The increase in diameter also allowed
shorter drums to be used, and the ratio of clearance allowance to
blade height was reduced, so as to avoid undue steam leakage. This
was rendered possible owing to the short drums being very stiff.
At the same time the weight of the turbine was reduced.

The other novel feature in the turbines for marine work was
that whereas in land turbines pistons were fitted on the rotor or
drum to balance the end pressure of the steam, the marine turbines
had dummies so proportioned as to leave an unbalanced end pressure,
which counteracted the thrusts of the propellers. There was,
therefore, no thrust bearing as with reciprocating engines. Thrust
or adjusting collars were fitted, but these were only to keep the
turbine in end adjustment and to take care of residual end pressure
at varying speeds.

Concurrently with the alterations in the machinery arrangements in the ship, experiments were made with model propellers
of various forms and proportions of projected blade area to disc
area. These test screw propellers were run by a weight and clockwork wheel train, a delicate spring balance registering the thrust.
The screw was mounted on an axis passing through a gland in
the side of a circular tin vessel 12 in. in diameter in such a position
that its thrust was tangential to the direction of rotation of the water
in the vessel. By this means the screw worked with a moderate
slip ratio in the rotating water. A plate-glass window was fixed
in the side of the vessel, through which the screw could be observed
and photographed. A view of this propeller-testing apparatus is
given on Plate XXXV. (Fig. 115).

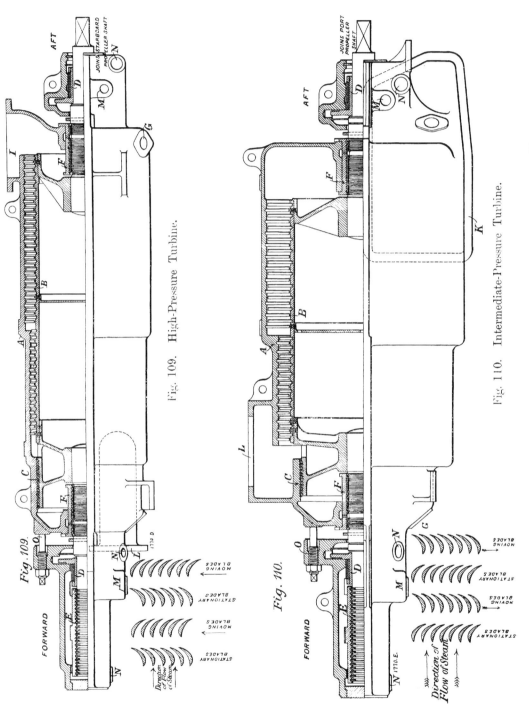

Fig. 109. High-Pressure Turbine.

Fig. 110. Intermediate-Pressure Turbine.

Plate XXXIII.—Parallel=Flow Turbines of the Second Installation of the "Turbinia."

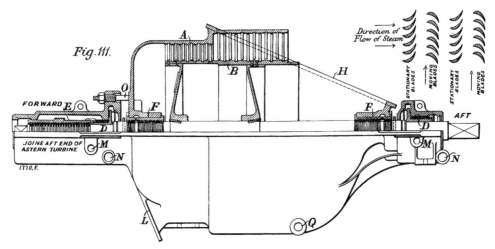

Fig. 111.   Low-Pressure Turbine.

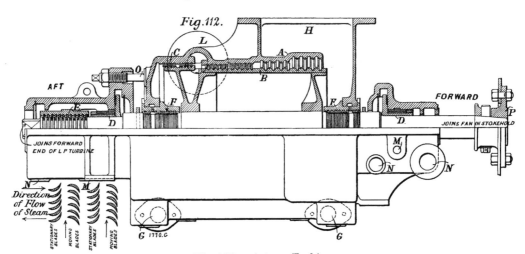

Fig. 112.   Astern Turbine.

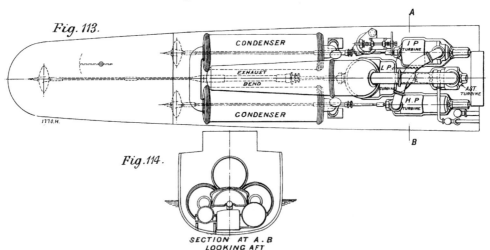

Figs. 113 and 114.   General Arrangement of Turbines.

Plate XXXIV.—Parallel-Flow Turbines of the Second Installation of the "Turbinia."

Fig. 115.   Propeller Testing Apparatus.

Fig. 116.   Beginning of Cavitation ; 1500 Revolutions.

Plate XXXV.—Experiments on Cavitation.

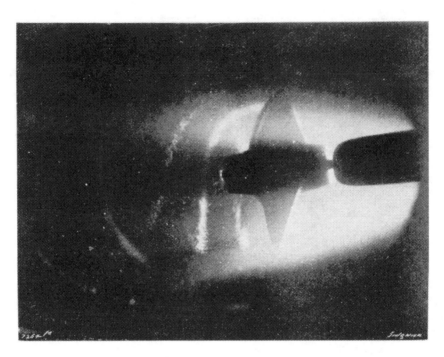

Fig. 117. Advanced Stage in Cavitation; 1800 Revolutions.

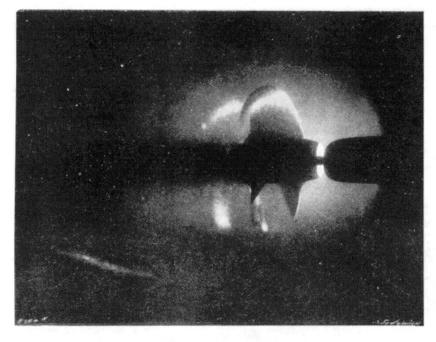

Fig. 118. Final Stage of Cavitation; 2000 Revolutions.

Plate XXXVI.—Experiments on Cavitation.

Photographs of the phenomena of propeller action were taken by intermittent illumination of the propeller from an arc lamp with an ordinary lantern condenser and mirrors, the propeller being thus illuminated in a definite position of each revolution for an exposure equal to about $\frac{1}{3000}$th part of a second; it therefore appeared stationary to the eye. With a propeller running at 1500 revolutions per minute and the water heated to nearly boiling point the cavities about the blades could be clearly seen and traced. It will be seen from the photographs given on Plates XXXV. and XXXVI. that a blister was first formed a little behind the leading edge and near the tip of the blade (Fig. 116); then, as the speed of revolution was increased, it enlarged in all directions, until at a speed corresponding to that of the " Turbinia's " original single propeller, it had grown so as to cover a section of the screw disc equal to 90 deg. (Fig. 117). When the speed was still further increased, the screw as a whole revolved in a cylindrical cavity, from one end of which the blade scraped off layers of solid water, delivering them to the other (Fig. 118). In this extreme case nearly the whole energy of the screw was expended in maintaining this vacuous space. This showed that when the cavity had grown to be a little larger than the width of the blade the leading edge acted like a wedge, the forward side of the edge giving negative thrust.

It was found when the blades of the propeller were broadened, so that the projected blade area reached about 0.7 of the disc area, the falling off of the thrust was very small, even in boiling water, while with a propeller of projected area (0.25 disc area) the thrust fell to a small fraction under the same conditions.

Another apparatus was then constructed consisting of an oval copper tank nearly full of water. Windows were placed on both sides, and, owing to the form adopted, the direction of flow of the water was more approximately parallel to the shaft than in the previous experiment. The air was withdrawn from the tank by a pump, so that there only remained to prevent cavitation the cohesion of the water and its " head " above the propeller. In tests of a 2-in. diameter propeller with wide blades cavitation

commenced at 1500 revolutions, the slip ratio being about 25 per
cent.   Much better observations and photographs could be made
than in the earlier apparatus, but thrust data were not taken with
this vacuum tank apparatus.   It has always been thought that if
the apparatus had been constructed on a scale ten times as large,
much valuable and accurate knowledge might have been obtained
regarding high-speed propellers, and such an apparatus is now in
course of construction at the Turbinia Works, Wallsend-on-Tyne.

As a result of these propeller experiments, trials of the
"Turbinia," with the new turbines, were resumed in February, 1896.
Three propellers, 18 in. in diameter and of one-pitch ratio, were
fitted on each shaft, making nine in all.   These gave better results ;
but it was decided to try several sets, designed as the outcome of
the experiments.   Nine propellers, of 18 in. diameter and 24 in.
pitch, were eventually adopted as apparently giving the best results,
the projected blade area being about 46 per cent. of the disc area.
Later experience, however, showed that, as will be explained later,
single propellers on each shaft are preferable in all ordinary types
of steamers.

The division of the power through three shafts tended to
increase propeller efficiency and speed.   An approximate estimate
was that the effective horse-power for a given volume of steam was
about doubled by the change.   Speeds exceeding 32 knots were
realised by the " Turbinia."   Such a result for a vessel only 100 ft.
long and of 44 tons displacement intensified the interest taken in
the marine turbine.   The three turbines, it was estimated, gave
2000 equivalent indicated horse-power—there was then no method
of measuring the power.   The turbines only weighed 3 tons 13 cwt.,
or, with boiler, shafting, propellers, &c., 22 tons.

During the following year, in April, 1897, tests were made of
the " Turbinia " by Professor Ewing, F.R.S., and one or two
quotations may here be taken from his report, published several
years later.[1]   Some twenty trial runs were made at speeds ranging
from $6\frac{3}{4}$ to $32\frac{3}{4}$ knots.   The steam consumption, power, and propeller slip, are given in the curves reproduced on Plate XXXVIII.

[1] "Proceedings of the Institution of Naval Architects," vol. xlv., page 293.

Plate XXXVII.—The "Turbinia" Steaming at 34 Knots (Fig. 119).

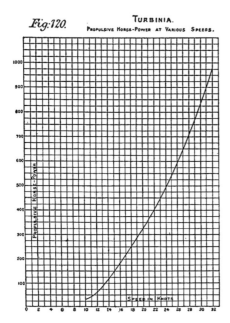

Fig:120.

TURBINIA.
PROPULSIVE HORSE-POWER AT VARIOUS SPEEDS.

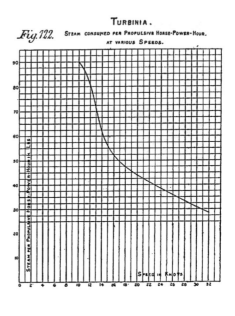

Fig.122.

TURBINIA.
STEAM CONSUMED PER PROPULSIVE HORSE-POWER-HOUR,
AT VARIOUS SPEEDS.

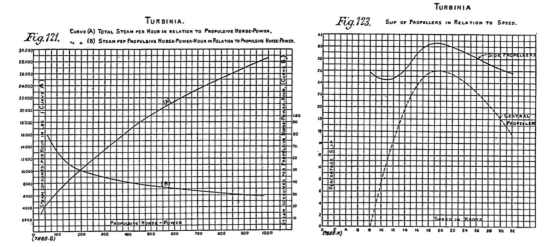

TURBINIA.
CURVE (A) TOTAL STEAM PER HOUR IN RELATION TO PROPULSIVE HORSE-POWER.
„ „ (B) STEAM PER PROPULSIVE HORSE-POWER-HOUR IN RELATION TO PROPULSIVE HORSE-POWER.

Fig.121.

TURBINIA
SLIP OF PROPELLERS IN RELATION TO SPEED.

Fig.123.

Plate XXXVIII.—Diagrams of Results of Trials of "Turbinia," by Professor Ewing.

The maximum speed was got with the wing propeller shafts making 2230 revolutions and the centre shaft 2000 revolutions per minute, the slips of the propellers being $25\frac{1}{2}$ and 17 per cent. respectively. The steam pressure at the boiler was 210 lb. and at the admission valve on the high-pressure turbine 157 lb. per square inch. The highest air pressure in the stokehold was $7\frac{1}{4}$ in. Referring to the diagrams of results, Professor Ewing says :—

"It is interesting to compare these results with those which are obtained in high-speed boats using engines of the ordinary type. The propulsive co-efficient or ratio of propulsive to indicated horse-power in such boats appears to be about 0.55 to 0.6, and their consumption of steam (at full power) is not in general less than 18 lb. per indicated horse-power per hour. Taking the most favourable co-efficient, this would correspond to 30 lb. of steam per propulsive horse-power hour as against the 29 lb. found in these trials of the 'Turbinia.' It is clear, then, that the exceptional speed developed in the 'Turbinia' has been achieved without sacrifice of economy, and that the substitution of turbines driving high-speed screws in place of reciprocating engines driving screws of much more moderate speed is not attended with increased consumption of steam so far as fast running is concerned.

"Taking the high-speed trials, we find in the turbine a rather smaller consumption of steam than is found in a high-speed engine for the same amount of propulsive effect. It is impossible to say definitely how the efficiency is distributed between the motors and the screws. I think it is likely that in the 'Turbinia' the motor is relatively rather more and the screws rather less efficient than in an ordinary boat. The propulsive co-efficient is probably not more than 0.5. Taking that value, the steam in the full-power trials was doing work in the turbines at the rate of about 2100 horse-power, and the consumption of steam was barely $14\frac{1}{2}$ lb. per horse-power hour.

"The engines of the 'Turbinia' develop approximately 50 horse-power per ton of displacement and 100 horse-power per ton of total weight, this total including the weight of engines, boiler, condenser, propellers, and shafts, tanks, water in boiler and hot well, and auxiliary engines.

"The turbines are remarkably handy, and allow sudden starts to be made with impunity.

"I have made experiments to see how rapidly speed could be got up. Starting from rest, a speed of rotation corresponding to 28 knots was acquired in 20 seconds after the signal was given to open the stop-valve. A conspicuous feature is the absence of vibration, due to the absence of reciprocating parts, and the perfect balance of the turbines. The contrast in this respect with ordinary high-speed boats is most striking. An incidental merit of the turbine is that it needs no internal lubrication, and the oil which circulates through the bearings has no opportunity of mixing with the steam. The advantage of having the condensed steam free of oil will be particularly felt in cases where water-tube boilers are used."

Throughout the long period over which these experimental trials of the "Turbinia" extended, the Admiralty had been kept informed by Mr. Parsons of the difficulties encountered and the means by which they were overcome. At the Naval Review of 1897 the "Turbinia" appeared, her moorings in Cowes roadstead being alongside Mr. C. J. Leyland's yacht, which was specially commissioned by him for the occasion. The extraordinary speed attained by her impressed upon all, especially upon naval officers, the possibilities connected with the use of the steam turbine. After preliminary negotiations, Mr. Parsons and his colleagues submitted to the Admiralty a proposal for building a turbine-driven destroyer, which should resemble closely in form and dimensions the 30-knot reciprocating-engined destroyers then building for the Royal Navy. This proposal was modified in details as the result of discussion between the professional officers of the Admiralty and Mr. Parsons, and in 1898 the "Viper" was ordered to be built, The Parsons Marine Steam Turbine Company being the principal contractors. With the sanction of the Admiralty, they placed the order for the hull and boilers of the vessel with Messrs. Hawthorn, Leslie, and Co. Complete responsibility for design, construction, and speed, however, rested upon The Parsons Marine Steam Turbine Company.

At the same time another vessel was built by Sir W. G. Armstrong, Whitworth, and Co., Limited, at their Elswick Works, under circumstances narrated in the introductory paragraph, and was later bought by the Admiralty and named the "Cobra." The "Viper" was 210 ft. long, 21 ft. beam, and displaced 370 tons; the "Cobra" was 223½ ft. long, 20½ ft. beam, and of about 390 tons displacement. The turbines in both vessels resembled in their general design those of the "Turbinia," but were arranged somewhat differently. Drawings of the machinery in the "Viper" are reproduced on Plate XXXIX., along with a view of the ship steaming at 37 knots. There were two sets of turbines, each with a high and low-pressure machine working in series, and placed respectively on the port and starboard side of the centre line of the ship. There were thus four shafts instead of three, as in the "Turbinia," each shaft being driven by one turbine, the outer by high-pressure and the inner by low-pressure machines. Each of the inner shafts had

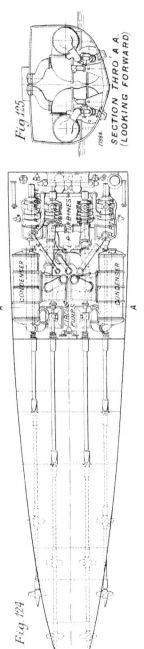

SECTION THRO' A.A.
(LOOKING FORWARD)

Fig. 125.

CONDENSER

H.P. TURBINES
L.P. TURBINES

CONDENSER

CIRCULATING
PUMPS

Fig. 124.

Figs. 124 and 125. General Arrangement of Machinery.

Fig. 126. Steaming at 37 Knots.

Plate XXXIX.—H.M. Torpedo Boat Destroyer "Viper."

an astern turbine in a separate casing at the forward end of the low-pressure ahead turbine. These astern turbines were connected with the condenser, and ran in vacuo when not in use. There were two screw propellers on each shaft, but later the "Cobra" had three screws on each shaft. The air-pumps were driven by separate small turbines with reduction gear. Provision was made during the trials for utilising the exhaust from the auxiliary engines in the main turbines. The more-recently adopted system of gearing and "closed exhaust" for auxiliary machinery had thus early recognition. Yarrow boilers were fitted in both ships, the heating surface being 15,000 square feet and the grate area 272 square feet.

Both vessels achieved remarkable results: the performance of the "Viper" may be reviewed. With full trial-weights on board and with a displacement of 370 tons, a mean speed of 36.581 knots was obtained on a one-hour's full-power trial. The fastest run on the measured mile was at the rate of 37.113 knots, and the mean speed in the fastest pair of runs 36.869 knots, the mean revolutions being 1180, and the mean air-pressure in the stokehold $4\frac{1}{2}$ in. The speed of 37.113 knots, it was estimated, required about 12,300 indicated horse-power for the 370 tons displacement weight, as compared with 6500 indicated horse-power developed in the 30-knot destroyers of similar dimensions, and of 310 tons displacement, built at the same time. At all speeds there was almost a complete absence of vibration. The astern speed was over $15\frac{1}{2}$ knots. The coal consumption, when steaming 31 knots ahead, was 2.38 lb. per equivalent indicated horse-power per hour. The mean speed of the "Cobra" on a three-hours' trial was 34.6 knots.

A thoroughly searching test was to have been made with the two vessels: but the unfortunate loss of both vessels made this impossible; the "Cobra" was lost in a storm in the North Sea on September 18, 1901, when on her way from the Tyne to the South, while the "Viper" was wrecked during manœuvres on the rocky coast of the Channel Islands, where she ran ashore during a fog on August 3, 1901. In neither case was the disaster in the remotest degree due to the turbines.[1] All the representatives of the

[1] See "ENGINEERING," vol. lxxii., page 452.

staff of the Parsons Company at Wallsend, of the builders, and many of the Admiralty crew, perished on the "Cobra."

Thus it came that after many years of careful work and experimenting, the "Turbinia" was the only vessel afloat with turbine engines, and in the public mind there was not the full realisation of the fact that the new mode of propulsion had nothing to do with the loss of the two destroyers. For a time the commercial prospects of the marine turbine were not brilliant. British business enterprise, however, asserted itself, and the application of the steam turbine to ship propulsion, though arrested, was not stopped. Thanks to the support of Mr. Christopher J. Leyland, a Director of the Parsons Company, and to the progressive spirit which has always been displayed by Messrs. Denny, Dumbarton, and by Captain John Williamson, long associated with Clyde steamers, the Parsons turbine was fitted in a new Clyde passenger boat—the "King Edward," as will be fully narrated in the next chapter. At the same time a torpedo-boat destroyer was laid down by The Parsons Marine Steam Turbine Company, Limited, with the munificent financial assistance of Mr. Leyland, who has throughout taken a very active part in the development of the turbine. This destroyer, the "Velox," was ultimately acquired by the British Admiralty.

The "Velox," launched in 1902 from the works of Messrs. Hawthorn, Leslie and Co., Limited, was of the same dimensions as the "Viper," and also had Yarrow boilers, the heating surface in this case being 13,200 square feet, and the grate area 252 square feet. The arrangement of the turbines corresponded to that in the "Viper" and "Cobra," there being four shafts, each with two screws, finally altered to one. The astern turbines were incorporated in the same casings as the low-pressure ahead turbines—now the usual practice. Small turbines were used for driving the forced-draught fans. The general arrangements and the machinery are shown on the same Plate as the engraving of the vessel steaming at 27.12 knots (Plate XL.).

As to the construction of the turbines in these earlier ships, the high-pressure drums carrying the blades consisted of lap-welded steel tubes. The low-pressure drums were weldless steel tubes. The wheels connecting the drums to the journal were of cast steel of the

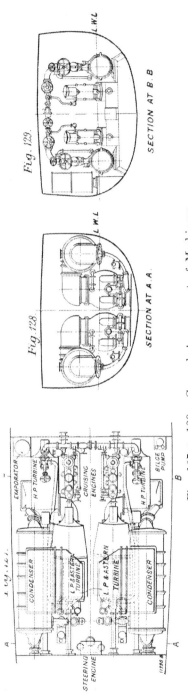

Fig. 129.

SECTION AT B. B.

Fig. 128.

SECTION AT A. A.

Figs. 127 to 129.   General Arrangement of Machinery.

EVAPORATOR

H.P. TURBINE

CRUISING
ENGINES

H.P. TURBINE

CONDENSER

L.P. & ASTERN
TURBINE

L.P. & ASTERN
TURBINE

CONDENSER

BILGE
PUMP

STEERING
ENGINE

Fig. 130.   Steaming at 27.12 Knots on Four Hours' Trial.

Plate XL.—H.M. Torpedo Boat Destroyer "Velox."

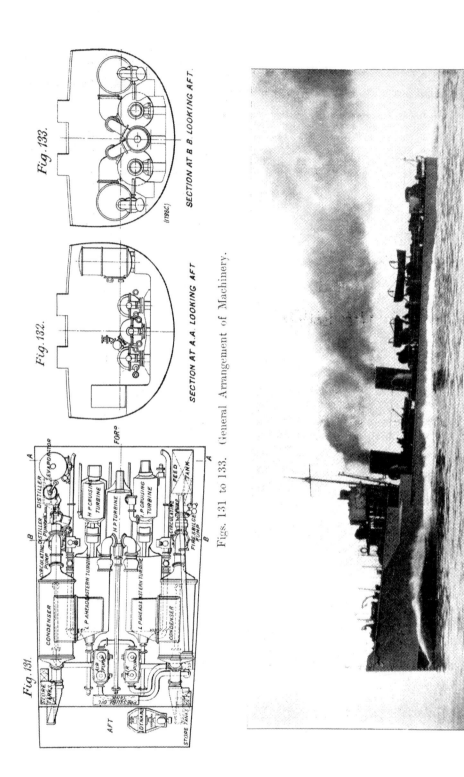

Fig. 133.

SECTION AT B B LOOKING AFT.

(1798C)

Fig. 132.

SECTION AT A.A. LOOKING AFT.

Figs. 131 to 133.   (General Arrangement of Machinery.

Fig. 131.

FOR.

DISTILLER   DISTILLER

CIRCULATING   EVAPORATOR
PUMP

H P CRUISING
TURBINE

H P TURBINE

L P CRUISING
TURBINE

FEED
TANK

B

A

CIRCULATING
PUMP

FIRE & BILGE
PUMP

L P AHEAD & ASTERN TURBINE

L P AHEAD & ASTERN TURBINE

CONDENSER

AIR
PUMP

CONDENSER

STORE
TANKS

AFT

PRESSURE OIL
TANK

HOTWELL

STORE TANKS

B

A

Fig. 134.   Steaming at 26.23 Knots on Four Hours' Trial.

Plate XLI.—H.M. Torpedo Boat Destroyer "Eden."

"arm" type. The dummies formed part of the drum or rim of the end wheel. The spindle or rotor shaft, of forged steel, was continuous through the drums. All scantlings were made as light as possible, consistent with strength and rigidity. The casings were of cast iron and the bearings of bearing metal lined with white metal. The blading, of the form shown in Figs. 66 to 70, on Plate XXV., facing page 46, was separately calked, with distance pieces, in grooves on drum and casing. The thrust or adjusting block also followed somewhat on the lines of the land practice, as already described in the preceding page.

Experience in the "Viper" and "Cobra" had shown that while the economy was at least as satisfactory at high speed as with reciprocating engines, efficiency fell off at low speeds. This is due to the inherent difficulty in turbines that the steam pressure falls directly with the quantity of steam passing, and the blade efficiency decreases with reduced revolutions. Inversely, it increases with high speed. Since warships do much of their steaming at low cruising speed, it became specially desirable to ensure economy at low power, and it was therefore decided to fit in the "Velox" two sets of triple-expansion reciprocating engines, each of 150 indicated horse-power, connected by detachable claw couplings, one on each of the inner shafts. These piston engines were for use only when steaming at speeds up to 12 or 13 knots, when they exhausted into the main turbines. At higher speeds they were disconnected at the couplings. This was the first application of the combination of reciprocating and turbine machinery extensively applied in low-speed ships, and dealt with in a subsequent chapter. The arrangement is shown on the plan on Plate XL., facing page 80.

The next destroyer, H.M.S. "Eden," ordered in the following year, and launched in 1903, was practically of the same dimensions as the "Velox." In the installation of turbines fitted in this ship, small turbines, for use only when cruising at low powers, were substituted for the reciprocating engines, as shown in the plan on Plate XLI. They were designed to give high efficiency at the low power required for cruising. One of these cruising turbines is of the high-pressure type, and the other of the low-pressure type, and

M

at cruising speed, after doing work in these successively, the steam passes through the ordinary high and low-pressure ahead turbines. For speeds up to about 14 knots both cruising turbines are used; between 14 and about 19 knots, the high-pressure cruising turbine is cut out and steam admitted direct to the low-pressure cruising turbine. For speeds above about 19 knots both cruising turbines are cut out and steam is admitted direct to the main high-pressure turbines. There is thus a great range of expansion, as well as a correct proportion of the blading and clearance areas for different speeds.

The Admiralty made careful tests of the results of trials of the " Velox " and " Eden " in comparison with corresponding tests of typical torpedo-boat destroyers driven by the best reciprocating machinery of the period—the culmination of a century's experience in piston engines. As shown in the Table on the next page, at 11 knots the water-consumption of the " Velox," with the reciprocating engines exhausting into the turbines was, on a three-hours' trial, 42 per cent. lower than that in the piston-engine destroyer. The " Eden," using cruising turbines exhausting into the main turbines on a four-hours' trial, required 9 per cent. more water per hour than the destroyer with the piston engines. The " Velox " thus established the gain with turbines using exhaust steam, a point to be dealt with later.

At 13 knots the " Velox," using only her high-pressure and low-pressure turbines, did not come out so satisfactorily, as the design did not anticipate their use at such low powers. The efficiency of the " Eden," still using both low-pressure and high-pressure cruising turbines, fell away slightly relative to that of the piston-engine ship. As speeds increased, the " Eden " steadily gained in relative economy, the water consumption at 15 knots being nearly 3 per cent. less, and at 18 knots 10 per cent. less, than in the ship with reciprocating engines.

On the specification four-hours' trial at the full speed of 25 knots, the " Eden" steamed at the rate of 3.39 nautical miles per ton of coal consumed, which was, at least, equal to the average performance of the destroyer class fitted with reciprocating engines. It may be added that the reciprocating engines in the "Velox" were replaced in 1906 by cruising high-pressure turbines to suit the increase in cruising speeds required by modern naval tactics.

The Admiralty had been considering the extension of the use of Parsons steam turbines to small swift cruisers during the period 1900-01, and in 1902 decided to adopt them in the "Amethyst," one of four third-class cruisers then building. In that vessel the arrangement of the turbines corresponded with that in the

TABLE VII.—WATER-CONSUMPTION PER HOUR AT VARIOUS CRUISING SPEEDS.

	Pounds per Hour.
At 11 knots.—"Velox," using piston engines exhausting into ordinary turbines	8,650
"Eden," using cruising turbines exhausting into ordinary turbines	16,146
Piston-engined destroyer	14,892
At 13 knots.—"Velox," using high-speed turbines only	24,375
"Eden," using cruising turbines exhausting into ordinary turbines	21,230
Piston-engined destroyer	18,140
At 15 knots.—"Velox," using high-speed turbines only	32,750
"Eden," using low-pressure cruising turbines exhausting into ordinary turbines	28,000
Piston-engined destroyer	28,750
At 18 knots.—"Eden," using low-pressure cruising turbine exhausting into high-speed turbines	41,050
Piston-engined destroyer	45,645

"Eden," and is illustrated by the general arrangement plan on Plate XLII., facing page 84. The high-pressure parts of the casings in both the "Eden" and "Amethyst" were made of cast steel, the remainder of cast iron. Great difficulty was experienced at this time in obtaining satisfactory steel castings, and one of the high-pressure turbine casings of the "Amethyst" developed a crack under steam when the cylinder was practically completed. In all designs subsequent to the "Amethyst" and "Eden" the casings have been made of cast iron. Generally, the scantlings of the casings and rotors were made slightly heavier than in the "Viper" and "Cobra."

As three other ships of exactly the same dimensions and form of hull as the "Amethyst" were ordered at the same time, and were fitted with piston engines designed by highly-experienced firms, a further opportunity presented itself for exact comparative trials.[1] The vessels, of 3000 tons displacement, were designed for a speed of $21\frac{3}{4}$ knots, with 10,000 indicated horse-power. The trials of the ships with reciprocating engines proved that the utmost that could be done with any degree of reliability was 22.34 knots ; the "Amethyst," for the same boiler-power, easily steamed at 23.63 knots, an increase in speed of 1.29 nautical miles per hour. But when it is noted that this gain was realised with easier steaming of the boilers, with the same weight of machinery, with no vibration of the ship—which enormously assists towards accuracy of gun-fire—and with quite 10 per cent. less coal per hour, and a proportionately greater radius of action, the superiority of the turbine will at once be appreciated. The absence of reciprocating parts reduces possibilities of wear and tear. The height required for the turbine machinery was 20 in. less, so that it could be more easily housed under a protective deck. The air-pressure in the stokehold was $\frac{1}{2}$ in. less, so that there was, as a result, less stress on the boilers. The manœuvring capabilities of the turbine cruiser proved quite as satisfactory as those of the reciprocating-engine ships. The time required for stopping from full-speed ahead or from starting from dead stop, ranged from $7\frac{1}{2}$ to 20 seconds, and only a few minutes were required to increase the ship's speed from 10 knots to 22 knots. The tactical advantage of the turbine could not therefore be doubted. A view of the ship at sea is reproduced on Plate XLIII.

Superior economy at all but the lowest speeds was clearly established by the exhaustive and carefully-conducted trials of all four ships. At low power, for 10-knots speed, the water consumption was about 23 per cent. higher in the "Amethyst" than in the reciprocating-engine ship. At 14 knots, the conditions were, so far as economy was concerned, more equal. When the speed was increased

[1] See "ENGINEERING," vol. lxxviii., page 689.

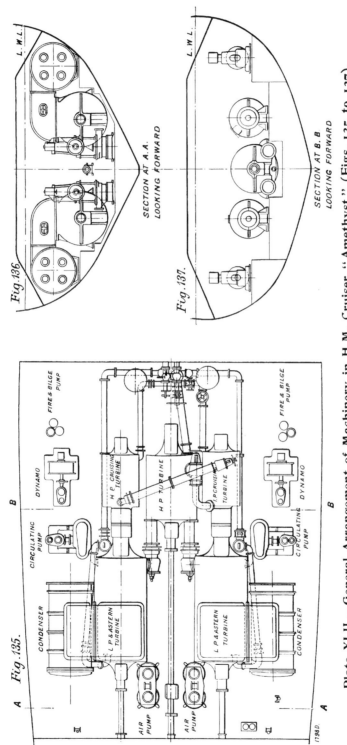

Fig. 136.

Fig. 137.

SECTION AT A.A.
LOOKING FORWARD

SECTION AT B.B
LOOKING FORWARD

Fig. 135.

L.P. & ASTERN TURBINE

L.P. & ASTERN TURBINE

CONDENSER

CONDENSER

CIRCULATING PUMP

CIRCULATING PUMP

DYNAMO

DYNAMO

FIRE & BILGE PUMP

FIRE & BILGE PUMP

H.P. CRUISING TURBINE

H.P. TURBINE

L.P. CRUISING TURBINE

AIR PUMP

AIR PUMP

Plate XLII.—General Arrangement of Machinery in H.M. Cruiser "Amethyst" (Figs. 135 to 137).

Plate XLIII.—H.M. Cruiser "Amethyst;" Speed attained, 23·63 Knots on Four Hours' Trial (Fig. 138).

to 18 knots, it was found that the consumption for the "Amethyst" was something like 20 per cent. less; at 20 knots it was nearly 30 per cent. less; and at the higher speed the improvement was still greater. The influence of this economy on the radius of action at high speeds is very marked; for instance, the turbine-propelled ship could, with her 750 tons of coal on board, steam 3160 sea miles at 20 knots, as compared with 2140 miles by the cruisers fitted with the ordinary machinery. After the trials provision was made for the utilisation of the auxiliary exhaust in the main turbines, which still further improved the economy at all speeds. The economy was as good as with the reciprocating-engined vessels at as low a speed as 14 knots.

The boilers in all the ships are of the "express" water-tube type, and it was specified that the coal burnt per square foot of heating surface per hour should not exceed 1 lb. In the "Amethyst" the Yarrow boiler was adopted, the fire-grate area being $493\frac{1}{2}$ square feet, and the heating surface 25,968 square feet, a ratio of 1 : 52.5. The air-pressures on the full-power trial in the reciprocating-engine ships ranged from 2.0 in. to 2.6 in.; in the "Amethyst," 1.9 in. on one trial, and 1.95 in. on another. The piston engines are of the four-cylinder triple-expansion type. The cruising high-pressure turbines of the "Amethyst" had 44-in. drums; the intermediate cruising turbine was of the same diameter, but with longer blades; the main high-pressure and low-pressure turbine drums were of 60-in. diameter; but the latter had longer blades.

The water-consumption results were arrived at by measuring the feed water in specially-constructed tanks. On the diagram, Fig. 139 on Plate XLIV., facing page 86, there is set out the water consumption per unit of power at various speeds in the respective ships. The power in the case of the "Amethyst" is, of course, assumed; but, as the form of the ships is identical in all cases, the same powers were taken, for comparative purposes, for the "Amethyst" as were required to drive the other ships at corresponding speeds. It is not, however, so easy to deduce the powers for the speeds attained by the "Amethyst" in excess of those realised by the others; but the projection of a progressive speed-curve affords an approximately accurate assumption. As, however, objection might be taken to this assumed power, the water-consumption per hour is plotted in Fig. 141 on Plate XLIV. for various speeds without regard to the power. It will be noticed that for lower speeds the turbine required more steam, but the "Amethyst" had not the advantage of the utilisation of the exhaust steam from the auxiliary machinery. In the "Topaze," for instance, when running at 10 knots, the water used for the auxiliary machinery was 4538 lb. per hour or over 21 per cent. of the total, and at 14 knots it was 5672 lb., or over 13 per cent. of the total. Although most of the auxiliary engines are compound, some are only simple expansion; the available heat in the steam exhausted from them is thus by no means a negligible quantity. In the reciprocating-engine ships this exhaust steam is led into the low-pressure cylinder receiver, whereas in the turbine ship it was passed into the condenser, although, as already stated, it was arranged later that the steam should pass to the turbines. The advantage of the turbine vessel in economy after 18 knots is obvious, and at 20 knots is very marked, while at the designed speed of $21\frac{3}{4}$ knots it is nearly 30 per cent.

The steam consumption of the "Topaze" shows a satisfactory thermal efficiency. Taking the water used for the main engines only, we find that at 18 knots the steam consumption was 15.45 lb., while at 14 knots it was 16.25 lb., and at 20 knots 16.91 lb. per horse-power per hour. In the "Amethyst" at 20 knots the steam consumption of the turbines, even including the auxiliary machinery, was about 14 lb. —a very exceptional figure. At 18 knots the water consumption was 15.5 lb., including that for auxiliary machinery, so that it would be a safe assumption to say that the main turbines were not taking more than 13 lb., which is probably a better result than has been realised by any high-speed naval reciprocating machinery in normal work. Disturbing variants come in—the weather, the skill of stokers, the calorific value of coal, &c.; but it is important to note, especially from the diagram, Fig. 140, showing the coal consumption per hour, that the coal results corroborate the water or steam economy of the turbine.

In the lower-power trials of both ships the evaporative results

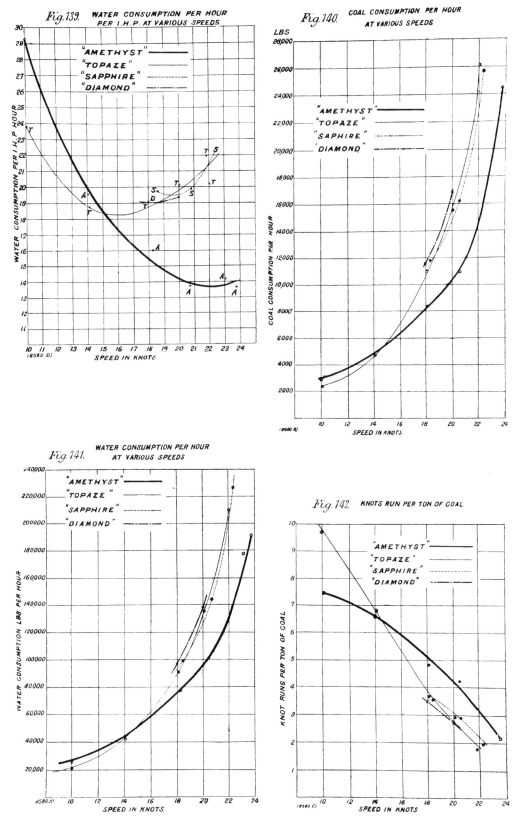

Fig. 139. WATER CONSUMPTION PER HOUR PER I.H.P. AT VARIOUS SPEEDS

"AMETHYST"
"TOPAZE"
"SAPPHIRE"
"DIAMOND"

Fig. 140. COAL CONSUMPTION PER HOUR AT VARIOUS SPEEDS

"AMETHYST"
"TOPAZE"
"SAPHIRE"
"DIAMOND"

Fig. 141. WATER CONSUMPTION PER HOUR AT VARIOUS SPEEDS

"AMETHYST"
"TOPAZE"
"SAPPHIRE"
"DIAMOND"

Fig. 142. KNOTS RUN PER TON OF COAL

"AMETHYST"
"TOPAZE"
"SAPPHIRE"
"DIAMOND"

Plate XLIV.—Diagram of Performances of "Amethyst" and of Sister Ships with Piston Machinery.

were fairly constant—about 9 lb. to 9¼ lb. per pound of coal; but with the pressing of the boilers to a greater extent, there was a slight falling off. In the case of the reciprocating-engined ships the boilers, when working at full power, were burning 50 lb. to 57 lb. per square foot of grate per hour, which, even with perfect control of the stokers, is not conducive to high evaporation; but even then the evaporation only slightly fell below 8 lb. in the "Topaze," and was 8.75 lb. in the "Sapphire."

A better measure of the coal economy is perhaps the distance travelled by the ships per ton of fuel at any given speed. At 18 knots the turbine-driven ship has an advantage equal to 30 per cent. over the "Topaze," and at 20 knots, of 45 per cent. The diagram, Fig. 142, showing the number of nautical miles run per ton of coal is thus interesting.

The propulsive efficiency is in all cases very favourable. In the "Topaze" four-bladed propellers were fitted on each of the two shafts, and in the "Sapphire" three-bladed screws; but the diameter, pitch, and area were in both cases exactly the same. The propellers on the "Amethyst" gave an average slip. In this case there was one propeller on each shaft, each of the propellers being three-bladed. The diameter in all cases was 6 ft. 8 in., but the pitch of the side screws was 5.91 ft., the area being 19.48 square feet, while the pitch of the centre propeller was 6.56 ft., with an area of 19.64 square feet. At 10 knots the mean slip of the three propellers works out to 11.3 per cent.; at both 14 and 18 knots, to 13.6 per cent.; at 20 knots, to 14.4 per cent.; at 23.06 knots, under very heavy weather conditions, 18.4 per cent.; and at 23.63 knots, in a smooth sea, to 17.1 per cent.

In the weight of machinery there is practically no difference between the turbine and reciprocating-engined ships, notwithstanding the fitting of cruising turbines in addition to the usual go-ahead and go-astern turbines. The total for the reciprocating-engined ships ranges from 530 to 540 tons, being in the case of the "Topaze" 537 tons; in the "Amethyst" the weights worked out to 535 tons. Assuming, therefore, that the machinery developed 14,000 indicated horse-power when the vessel made 23.63 knots, it is evident that the power developed per unit of weight was 26 indicated horse-power per ton; whereas in the "Topaze" it was only 18.3 indicated horse-power per ton.

There is justification for so fully recording the results of the comparative trials of these vessels, because they had a great influence upon the future of the turbine. They established the efficiency of the system beyond all doubt, and it was decided to adopt turbines for all torpedo craft. Indeed, it was soon recognised that without them the higher speeds since achieved, in combination with superior fighting qualities and radius of action, would have been impossible. Moreover, when a committee, representative of the service and of all branches of science, was appointed by the Admiralty in 1905 to consider the question of the design of armoured ships, they recommended the adoption of the system for all such ships. The battleship "Dreadnought," the design of which was accepted by this committee after full consideration, marked a signal departure.

The turbines for this ship were made at the Wallsend Works of the Parsons Company, under sub-contract with Messrs. Vickers Sons and Maxim, Limited. They embodied the improvements in detail which had been made from time to time in the machinery of the successive merchant and naval ships, some of which will fall to be considered in the next chapter dealing with work for the mercantile marine. The three-shaft arrangement was preferred for relatively small passenger steamers, but for several reasons it was deemed desirable in naval practice to have four shafts for the large ships, so that the machinery on each side of the centre line could be independent, with a water-tight bulkhead separating the two engine rooms. On each outer shaft there is a high-pressure ahead and a high-pressure astern turbine in separate casings; and on each inner shaft a cruising turbine, and low-pressure ahead and astern turbines in one casing. The two ahead turbines and the two astern turbines in each engine room work in series respectively. This arrangement of machinery for a battleship is illustrated in the drawings reproduced on Plate XLV.

Careful consideration was given to general accessibility for

HALF SECTION AT "B" LOOKING FORD   HALF SECTION AT "A" LOOKING AFT.

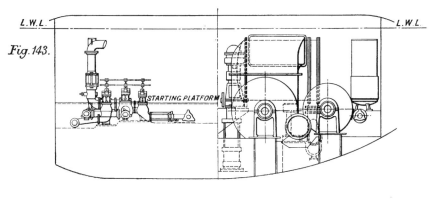

Fig. 143.

L.W.L.   L.W.L.

STARTING PLATFORM

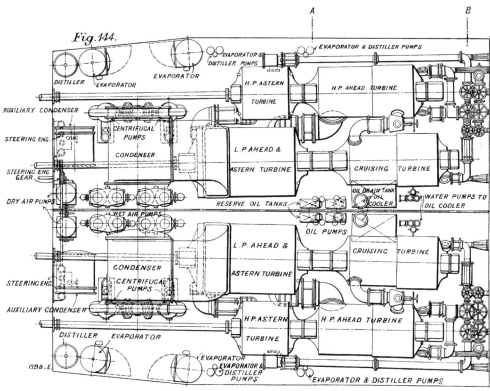

Fig. 144.

A   B

EVAPORATOR & DISTILLER PUMPS
EVAPORATOR &
DISTILLER PUMPS
DISTILLER   EVAPORATOR   EVAPORATOR
AUXILIARY CONDENSER
H.P. ASTERN
TURBINE   H.P. AHEAD TURBINE
CENTRIFUGAL
PUMPS
STEERING ENG.
CONDENSER
L.P. AHEAD &
STEERING ENG.   CRUISING TURBINE
GEAR   ASTERN TURBINE
DRY AIR PUMPS
OIL DRAIN TANK
OIL
COOLER   WATER PUMPS TO
WET AIR PUMPS   RESERVE OIL TANKS   OIL COOLER
OIL PUMPS
L.P. AHEAD &   CRUISING TURBINE
CONDENSER
ASTERN TURBINE
CENTRIFUGAL
STEERING ENG.   PUMPS
AUXILIARY CONDENSER
H.P. ASTERN
DISTILLER   EVAPORATOR   TURBINE   H.P. AHEAD TURBINE
EVAPORATOR
1198.E   EVAPORATOR &
DISTILLER
PUMPS
EVAPORATOR & DISTILLER PUMPS

Plate XLV.—Turbine Machinery of a Battleship.

Plate XLVI.—H.M. Battleship "Dreadnought" (Fig. 145).

overhauling. The rotor drums were hollow steel forgings built up with cast-steel wheels and spindles as in previous practice, with the exception that the steam end wheels were made with hollow arms. It had been observed in one or two previous vessels on service that the dummy clearance varied somewhat owing to the difference in temperature under varying conditions. With a view to maintaining as far as possible a uniform clearance at all temperatures at the ahead dummies of the contact type, the arms of the steam end wheel which connects the spindle to the drum have, in most of the turbines constructed since the turbines of the " Dreadnought," been made hollow, to allow steam to be admitted to the shaft. The shaft was also made hollow for a certain length to equalise any differences in expansion arising from unequal temperatures between the spindle and the casing. The steam-packed glands, where the turbine shaft passes through the cylinder casing, were, in the case of the turbines of the " Dreadnought " and subsequent ships, made quite independently of the main cylinder in order to permit of examination of the gland rings without interfering with the main turbine covers. Prior to the building of the " Dreadnought's " turbines, a loose coupling was fitted to connect the turbine to the line shafting, but in this and subsequent machines the rotor shafting was slightly increased in diameter, thereby increasing the size of the gland rings which permitted them to be worked over a solid coupling where the turbine rotor was connected with the shafting.

The turbine casings were arranged in parts with vertical flanges connected by bolting, and the ends of the turbines were " coned " inward for additional strength to resist end pressure. Large oil wells embodied in the bearing ends were introduced at this time, and into these the heated oil flowed from the bearings and adjusting blocks, the object being to minimise differences in temperature of the various parts of the turbines in order to obtain, as far as possible, uniform dummy clearance.

In the earlier vessels, H.M.SS. " Viper," " Cobra," and " Velox," and all the vessels prior to H.M.S. " Eden," the necessary adjusting block alterations were carried out by means of portable levers. From the date of H.M.S. " Eden," and until 1906, a special bracket

N

and screw were fitted on the forward end of the cylinder-bearing block, and by means of this screw and suitable gear the rotors could be adjusted in the fore and aft direction as desired. A slight modification to this gear was adopted in 1906 to enable the rotors to revolve more readily and under complete control, and this form of gear is now commonly adopted.

In the original design of Parsons turbine the longitudinal adjustment and clearance of the dummies were regulated by the position of a finger piece fixed on the turbine shaft just outside the cylinder casing at the forward end; but it was thought that some arrangement should be made to enable the clearance of the dummies to be correctly checked inside the cylinder. About 1906 Messrs. Denny, of Dumbarton, introduced the micrometer gauge for direct measurement of the dummy clearances, and in H.M.S. "Dreadnought" and in subsequent designs such micrometer gauge was adopted. By it the clearance can be measured and regulated, if necessary, almost to one-thousandth of an inch.

From 1907 there has been fitted in all large vessels a special type of permanent adjustment gear in conjunction with the adjusting blocks, whereas formerly the adjusting gears were portable, and the keeps had to be removed to be applied. This gear is well shown in the photographs of the turbines of the Japanese liner "Tenyo Maru" reproduced on Plate XCVII., in a later chapter dealing with turbines in foreign ships.

In supplement to this brief review of the mechanical improvements in warship turbines, we may quote, as an authoritative summary of the advantages of the system, the reasons given by the First Lord of the British Admiralty for the adoption of the Parsons turbine in the "Dreadnought" and succeeding ships as determined by a committee on naval design :—

"The question of the best type of propelling machinery to be fitted was most thoroughly considered. While recognising that the steam-turbine system of propulsion has at present some disadvantages, yet it was determined to adopt it because of the saving in weight and reduction in number of working parts, and reduced liability to breakdown, its smooth working, ease of manipulation, saving in coal consumption at high powers, and hence boiler-room space, and saving in engine-room complement; and also because of the increased protection provided for with this system, due to the engines being lower in the ship—advantages which much more than counterbalance the disadvantages. There was no difficulty in arriving at a decision to adopt turbine propulsion from the point of view of sea-going speed only. The point that chiefly occupied the Committee was the question of providing sufficient stopping and turning power for purposes of quick and easy manœuvring. Trials were carried out between the sister vessels 'Eden' and 'Waveney' and the 'Amethyst' and 'Sapphire,' one of each class fitted with reciprocating and the other with turbine engines; experiments were also carried out at the Admiralty Experimental Works at Haslar, and it was considered that all requirements promise to be fully met by the adoption of suitable turbine machinery, and that the manœuvring capabilities of the ship when in company with a fleet, or when working in narrow waters, will be quite satisfactory."

Sir Henry Oram, the Engineer-in-Chief of the Fleet, in his address as president of the Junior Institution of Engineers,[1] stated that, at full power, the steam consumption of the "Dreadnought" was 13.48 lb. per shaft horse-power per hour, while in the succeeding battleships of the class it averaged 13.01 lb., and in the three cruisers of the Invincible class 12.03 lb. With reciprocating engines nearly 16 lb. would be a fair average, and it thus follows that a great reduction in boiler weights was permissible. Again the high efficiency of the low-pressure turbine made it well worth while to pass the exhaust steam from the auxiliary engines to this turbine instead of to the condenser. Indeed, the exhaust steam in some battleships is sufficient alone to drive the vessel at a speed of five to six knots.

The coal consumption at full power of the three 26-knot armoured cruisers of the Invincible class ranged from 1.2 lb. to 1.7 lb. per shaft horse-power per hour, the average for the three ships being 1.47 lb. per shaft horse-power. In the three cruisers of the Minotaur class, with piston engines, it was 1.8 lb., and in the six cruisers of the Duke of Edinburgh or Warrior class 2.1 lb. per indicated horse-power per hour. On the thirty-hours' endurance trial, at 70 per cent. of the total power, the turbines also proved more efficient, although the advantage was not so marked.

At one-fifth power the coal consumption of the three Invincible cruisers averaged 2.4 lb. per shaft horse-power per hour, as compared with 1.87 lb. per indicated horse-power per hour in the Minotaurs and 2.05 lb. in the Duke of Edinburgh cruisers.

[1] See "ENGINEERING," vol. lxxxviii., page 703.

On this low-power trial the cruising turbines were, of course, in use in ships of the Invincible class. In the British service the navigating authorities insist on having the turbines on each side of the centre line independent, whereas in some other services, notably the German, the two cruising turbines, one on each side of the centre line, are worked in series, the port cruising turbine exhausting into that on the starboard side, or *vice versâ*. This is conducive to higher economy at low power, but has the objection that there must be a steam pipe and gland through the centre line bulkhead. Now in larger warships, instead of having separate cruising turbines, the main turbines are increased in dimensions partly to compensate for the loss of economy at low powers by the omission of the cruising turbine, and to accommodate an extra stage of shorter blades at the initial end of the high-pressure turbine as a "cruising element." At full power additional steam is by-passed to the second stage of the high-pressure turbine.

There is thus not only economy of steam but reduction in weight of machinery. A comparison of the reciprocating and turbine machinery of two typical recently completed warships of about the same brake horse-power bears out this fact. The steam generating capacity of the boilers in the turbine-driven ship is 20 per cent. less than in the reciprocating-engined ship, and the total weight of machinery is 10 per cent. less.

Again, it has been stated that the boilers of the cruisers of the Invincible class would require to have been 40 per cent. heavier to enable 43,000 shaft horse-power to be developed under the same degree of forcing. The consequent addition to the displacement of the ship, presuming other fighting elements to have been the same, would have made a speed of 26 knots practically impossible.

In destroyers the weights of the turbine and reciprocating machinery do not differ much, but the limit in speed with reciprocating engines was reached in 30-knot boats. In the design of these vessels the first consideration is speed; machinery weight and fuel consumption are secondary. With reciprocating engines, running at 400 revolutions per minute, it was, as one engineer put

Plate XLVII.—H.M. Cruiser "Indomitable" (Fig. 146).

This Warship conveyed His Royal Highness the Prince of Wales (now His Majesty King George V.) to Canada, for the Centenary Celebrations at Quebec in 1908, and on the return voyage averaged 24.8 knots for the Transatlantic voyage, steaming for 72 nautical miles at a mean speed of 25.13 knots (see "Engineering," vol. lxxxvi, page 178).

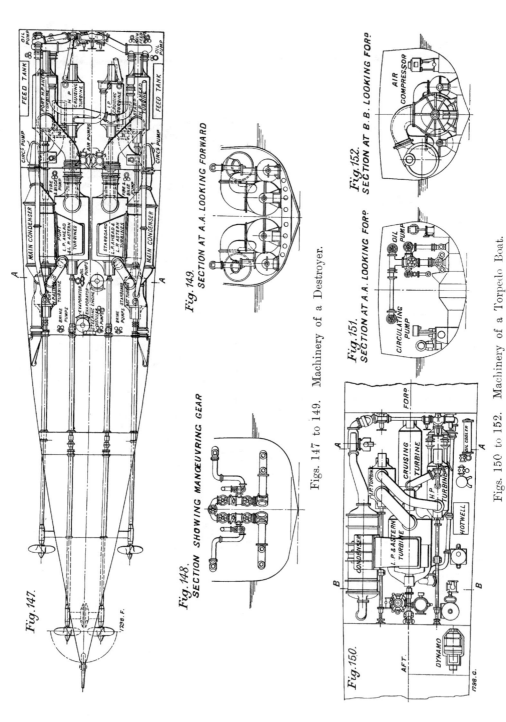

Fig. 147.

Fig. 148.
SECTION SHOWING MANŒUVRING GEAR

Fig. 149.
SECTION AT A.A. LOOKING FORWARD

Figs. 147 to 149.   Machinery of a Destroyer.

Fig. 150.

Fig. 151.
SECTION AT A.A. LOOKING FOR?

Fig. 152.
SECTION AT B.B. LOOKING FOR?

Figs. 150 to 152.   Machinery of a Torpedo Boat.

Plate XLVIII.—Machinery of Later Torpedo Craft.

it, a case of "pouring in oil, and trusting to God." The engineer-contractor of some of the latest torpedo-boat destroyers has put it well :—" With reciprocating engines having the weights cut down to a minimum most careful inspection is required to ensure that the piston and connecting rods, bolts, and the like, are sound. On rough turning small markings can be detected in these by the experienced eye, which indicate not perfectly satisfactory metal. The finishing tool will, however, cover up faults one would have liked to have seen, and consequently on trial one is always asking oneself the question : 'Have I found all the bad ones'? A failure at full speed means the loss of all in the engine room. With the turbine, on the other hand, there is absolute peace of mind. The men, indeed, complain that in long trials they cannot keep awake, as they have nothing to do."[1]

On Plate XLVIII. there is illustrated the general arrangement of the turbine machinery in a high-speed torpedo-boat destroyer with four shafts, and in a torpedo-boat with three shafts. In the latter case a large part of the expansion—about one-half of the whole power—is put on the low-pressure turbine on the centre shaft, which runs at a slower speed. As the centre shaft alone has an astern turbine, this arrangement gives a greater power for reversing.

H.M.S. "Swift," which is illustrated on Plate XLIX., is the largest torpedo-boat destroyer yet built. The arrangement of machinery is generally similar to that shown on Figs. 147 to 149 on Plate XLVIII. Although the vessel is only 345 ft. long and 34 ft. 2 in. beam, with a displacement tonnage of 2170 tons, it was found possible to accommodate, alike as regards weight and space, turbines which developed on some of the experimental trials nearly 50,000 shaft horse-power. On her official trials she steamed 35.3 knots with the turbines developing 35,000 shaft horse-power. The vessel was constructed by Messrs. Cammell Laird and Co., Limited, of Birkenhead.

A typical modern British destroyer is illustrated on Plate L. This vessel, the "Mohawk," was constructed by Messrs. J. S. White and Co., Cowes. With a length of 270 ft., and a beam of 25 ft.,

[1] At the Conference of the Institution of Civil Engineers, 1907.

she displaces 865 tons at 8.9 ft. draught. The turbines drive three shafts, and are arranged generally as in Figs. 150 to 152 on Plate XLVIII. When developing 14,500 shaft horse-power they gave the vessel the very satisfactory speed of 34.51 knots.[1]

Experience has greatly lessened the objection formerly taken to the non-reversibility of the turbine. In some of the earlier ships the astern turbines were, perhaps, of insufficient power, but now they are, as a rule, larger, and take the full volume of steam that the boilers give, developing a power equal to 50 per cent. of the maximum of the go-ahead turbines. In the turbine the power to stop or to reverse the direction of the ship is more effective than with piston engines, because the reversing turbine at the beginning of its work is still rotating in the ahead direction by reason of its momentum, and therefore against the flow of steam for going astern; the effective steam pressure on its blades is very considerably greater than when running normally, and thus the stopping or brake power is very much greater than would otherwise be the case. Indeed, this stopping or brake power immediately after reversing may be 40 per cent. to 60 per cent. higher than the normal astern going power, and it is at the moment of reversing that the greatest brake power is needed. No collisions have occurred owing to the deficient astern power of turbines; vessels with the turbine machinery running at 10 knots have been brought up in from one to one and a-half times their length, and at 19 or 20 knots from two to two and a-half times their length.[2] Turbine-driven vessels have steamed astern at from 60 per cent. to 70 per cent. of their full speed ahead. Another important factor is the time in which the direction of rotation of the shaft can be changed; it has been done in turbine steamers within twelve seconds of the giving of the order from the bridge.

Of greater importance in manœuvring is the turning circle of the ship, and lately there has been considerable increase in the area of rudder fitted, and consequently a reduction in the turning

[1] See "ENGINEERING," vol. lxxxiv., page 722.
[2] See "Transactions of the Institution of Naval Architects," vol. xlix., page 272.

Plate XLIX.—H.M. Destroyer "*Swift*"; *Speed attained on Official Trial*, 35.3 Knots (Fig. 153).

Plate L.—H.M. Destroyer "Mohawk"; Speed attained, 34·51 Knots (Fig. 154).

circle of the ships. The diameter of the turning circle has, in some instances, been reduced to two and a-half times the length of the ship, and, except in very restricted harbours, it would seem to be much easier to avoid collision by giving the ship helm, while maintaining the speed, than by reversing the engines.

Thus has the efficiency of the turbine for warship propulsion been demonstrated by successive warships, and step by step naval opinion has been won to a universal acceptance of the confidence of early workers. Of this there is evidence in the advance within twelve years from the 2000 horse-power of the "Turbinia" to the unprecedented powers of the latest fighting leviathans, a triumph which has been well merited as a consequence of the untiring work in research done not only in perfecting mechanical details but in order to overcome the difficulty of reconciling the necessary high speed of the turbine and the equally important low speed of the propeller in order to secure combined efficiency.

THE DEVELOPMENT OF THE TURBINE IN THE BRITISH MERCHANT SERVICE.

THE progress in turbine machinery in the merchant service has been as satisfactory as in the Royal Navy. The first mercantile ship fitted with the system was completed as recently as 1901, and within seven years the horse-power in one steamer increased from 3500 to 78,000—the mean power developed on a 1250-mile trial of the Cunard liner "Mauretania." In the design of the machinery for the first merchant ship there was available the experience gained in the destroyers which preceded it, and, as a consequence, a high economy was realised, the steam consumption for the main engines only being about $14\frac{1}{2}$ lb. per shaft horse-power per hour. In the succeeding seven years the rate of consumption was reduced to $11\frac{1}{2}$ lb.

The results so far achieved have eminently justified the pioneers in the application of the new system to the mercantile marine. But that the first step was, at the time, a serious undertaking is indicated by the extreme reluctance of merchant shipowners to embark upon what seemed a costly experiment, notwithstanding the guarantees of success which had been afforded by the performances of the "Turbinia," the "Cobra," and the "Viper." It is true that the conditions obtaining in the Naval service differ essentially from those in the mercantile marine. In the Navy, efficiency is the first consideration, rather than low first cost and working charges; again in the merchant service the rate of speed is practically constant, whereas, in the fighting fleet, speed must vary to suit the ever-changing demands of war tactics or peace manœuvres. The machinery for the merchant ship may therefore be designed to give its highest efficiency at the fixed service speed. In the Navy, on the other hand, efficiency is desirable at all rates of speed. This essential difference in service conditions makes the problem of the application and design of turbines for fast passenger ships more simple. In recent ships, where the practice has been adopted of requiring the builder to guarantee only the service speed, rather than a rate in excess of it in order to allow for emergency, the turbines have been designed to give their maximum efficiency at the service speed, any extra power to make up lost time being provided by temporary overload.

In the earlier vessels the steam consumption, when developing a small fraction of full power, was higher than with piston engines, for reasons indicated in preceding chapters; but at or near full power the results were at least as good as with reciprocating engines. Other considerations, in addition to economy, justified the adoption of the turbine in warships—some of these have already been referred to in the preceding chapter—but the dominant idea in the mind of the shipowner is exclusively associated with economy, and there was some hesitancy to accept the satisfactory results at high powers in early warship trials, associated as they were with a lack of economy at low powers.

Mr. Parsons from the first maintained the opinion that the steam turbines would prove superior to the reciprocating engine in mercantile steamships, as well as warships, and illustrated these advantages in papers read before engineering and scientific societies. Amongst these may be mentioned a paper contributed to the Institute of Marine Engineers in 1897, in which it was stated that "in applying turbine engines to a large passenger vessel of, say, 30,000 indicated horse-power, probably four screw shafts, with two screws on each shaft, would be adopted. . . . With such engines the consumption of steam per propulsive horse-power would probably be less than that found in the mercantile marine. . . . . With turbine engines in passenger vessels there would arise no questions of vibration from machinery or propellers, and in the event of one screw shaft, or one motor, becoming disabled the one affected can be more readily taken out of action than is the case with ordinary engines and . . . thus the liability to serious break-down is considerably reduced." It is unnecessary to comment on this remarkable anticipation of what has since become current practice in the swiftest passenger steamers on service.

Again, at the meetings of the British Association at Dover in 1899 Mr. Parsons (at the request of the President of Section G—Sir

o

William White) contributed to the proceedings a valuable paper dealing
in detail with the application of steam turbines to the propulsion of
swift Cross-Channel and coasting steamers, and described arrangements
which, in their main features, anticipated present practice. In January,
1900, on the occasion of an evening discourse, delivered at the Royal
Institution, Mr. Parsons reiterated his conviction that " in the case of
an Atlantic liner . . . turbine engines would effect a reduction in
weight of machinery and an increased economy of fuel." Many other
examples might be given of this advocacy of the new system of propulsion and confidence in its complete success when there were few who
shared the opinions of Mr. Parsons, and when no shipowner was bold
enough to make the experiment.

Much persuasion availed nothing, and it was ultimately
recognised that the only chance of convincing shipowners was for
Mr. Parsons and his friends to undertake the financial responsibility
of demonstrating the results by fitting the system in a merchant
vessel. Negotiations were almost completed for the purchase of an
old paddle steamer in which to fit a set of experimental turbines;
this idea was, however, departed from, and that fortunately, because
the lines of the steamer might have very seriously interfered with
the propulsive efficiency. The change in procedure was due to
Messrs. Denny, of Dumbarton—on the initiation of Mr. Archibald
Denny of that firm—taking a more liberal view of the situation
than some of their *confrères* in the shipbuilding industry. At the
same time Captain John Williamson, who had been so long identified
with the Clyde passenger paddle steamers, also showed great enterprise in recognising the potentialities of the system and fell in with
the idea of running a turbine steamer on the Clyde Estuary. Three
of the partners of the Denny firms—Mr. James Denny, Mr. John
Ward, and the late Mr. Henry Brock—attended trials of H.M.S.
" Viper," and as their report, backed by Mr. Archibald Denny,
was highly favourable, the result was the building, in 1901, by
Messrs. Denny, of the " King Edward," the first merchant turbine
steamer, in the ownership of which Captain John Williamson,
Messrs. Denny, and the Parsons Company combined. Mr. Christopher Leyland, of Haggerston Castle, Northumberland, who had

Plate LI.—The First Turbine Merchant Steamer, "King Edward" (Fig. 155).
(Speed attained, 20.48 Knots.)

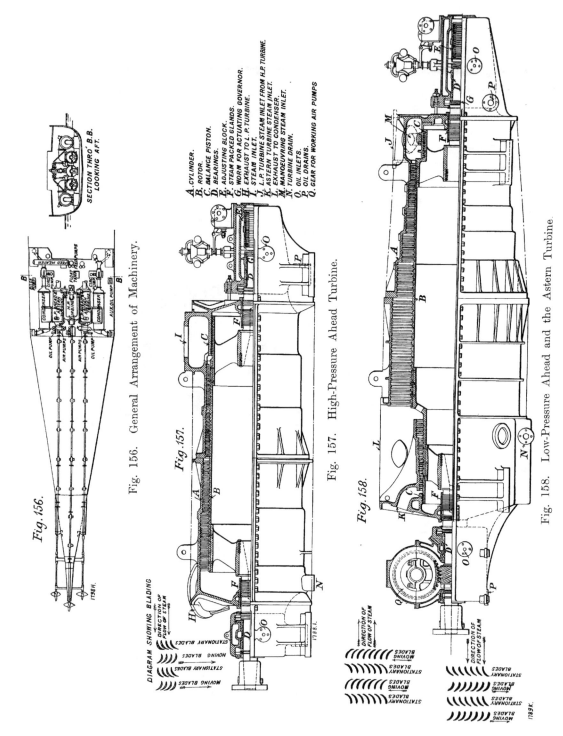

Fig. 156.

SECTION THRO' B.B.
LOOKING AFT.

Fig. 156. General Arrangement of Machinery.

DIAGRAM SHOWING BLADING

DIRECTION OF FLOW OF STEAM

STATIONARY BLADES
MOVING BLADES
STATIONARY BLADES
MOVING BLADES

A. CYLINDER.
B. ROTOR.
C. BALANCE PISTON.
D. BEARINGS.
E. ADJUSTING BLOCK.
F. STEAM PACKED GLANDS.
G. WORM FOR ACTUATING GOVERNOR.
H. EXHAUST TO L.P. TURBINE.
I. STEAM INLET.
J. L.P. TURBINE STEAM INLET FROM H.P. TURBINE.
K. ASTERN TURBINE STEAM INLET.
L. EXHAUST TO CONDENSER.
M. MANŒUVRING STEAM INLET.
N. TURBINE DRAIN.
O. OIL INLETS.
P. OIL DRAINS.
Q. GEAR FOR WORKING AIR PUMPS

Fig. 157.

Fig. 157. High-Pressure Ahead Turbine.

Fig. 158.

DIRECTION OF FLOW OF STEAM
STATIONARY BLADES
MOVING BLADES
STATIONARY BLADES
MOVING BLADES

DIRECTION OF FLOW OF STEAM
STATIONARY BLADES
MOVING BLADES
STATIONARY BLADES
MOVING BLADES

Fig. 158. Low-Pressure Ahead and the Astern Turbine.

Plate LII.—The Turbine Machinery of the "King Edward," the First Turbine Merchant Ship.

been actively interested in the development of the marine steam turbine from the beginning, took a prominent part in this new departure and did much to carry it to success.

The "King Edward," of which an engraving is given on Plate LI., is a vessel 250 ft. in length, 30 ft. in breadth, and at 6 ft. draught has a displacement tonnage of 650 tons. She was fitted by The Parsons Marine Steam Turbine Company with turbines working three shafts, as shown in the general arrangement, Fig. 156, on Plate LII., facing this page. There is one high-pressure turbine on the centre shaft (Fig. 157), and in this steam is expanded five-fold; on each wing there is a low-pressure turbine (Fig. 158). Both take steam from the high-pressure turbine, and expand it 25-fold, giving a total expansion of 125-fold. The astern turbines are incorporated in the same casing as the low-pressure ahead turbines (Fig. 158). The two wing shafts, which are therefore used for astern as well as ahead driving, had originally two propellers, while the centre shaft had a single propeller. Later there was fitted a single propeller to each shaft, as shown in the plan, Fig. 156. The air and oil pumps are worked off the low-pressure turbine shafts through worm gearing. The turbines differed in several respects from those then manufactured for destroyers. The cylinders were of somewhat heavier construction, and were made of cast iron. The question of greater accessibility to parts was considered. The rotors were constructed without the shafts being carried right through the drum, the turbine-shaft ends being shrunk into the wheels carrying the drums. The drums in this vessel were also of steel, and lap-welded, and the wheels were made of cast steel of the "armed" type.

The "King Edward" conducted a short service in the estuary of the Clyde, involving 150 to 180 miles' steaming per day, and as there were paddle vessels engaged in the same duty direct comparison was possible. These paddle steamers were probably the most efficient of the type constructed, as a consequence of the long experience of the Clyde shipbuilders, so that when the "King Edward" established a distinct superiority some progress was made in the education of the shipowner in favour of the system. The results attained by the "King Edward" undoubtedly marked a distinct step in the development of the turbine, and the name of Captain John Williamson, who was partly responsible for the construction, and entirely for the management, of the ship, will long be identified with the system of propulsion as the owner of the pioneer turbine merchant ship.

The success of the "King Edward" resulted in a second vessel being built for the same service in the following year, 1902. This vessel, of which an engraving is given on Plate LIII., facing this page, is named "Queen Alexandra," and is slightly larger, being 270 ft. long and 32 ft. beam, and at 7 ft. draught displaces 750 tons. She carries 2077 passengers, as compared with 1994 in the case of the "King Edward," and her service speed is 19 knots, as compared with $18\frac{1}{2}$ knots. The turbine machinery is very similar to that in the "King Edward," but the air and oil pumps are independently driven, the air and circulating pumps being worked by one engine.

In service during the season 1904 these two vessels not only did a much greater mileage, but had a speed of 2 to 3 miles per hour higher, than the paddle vessels, and burnt less fuel. The coal consumption over a period of years has proved to be nearly 20 per cent. lower than in vessels of similar dimensions and speed, but fitted with compound engines driving paddle wheels.

The sustained success of the "King Edward" and "Queen Alexandra" naturally attracted the attention of the railway companies conducting services between the British Isles and Continental ports, as well as between Britain and the Irish ports, as the conditions of service—full speed for the complete voyage—were even more favourable owing to the absence of intermediate stops. Thus the two Clyde vessels became centres of keen observation, and they rarely made a trip without having on board officers from shipowning companies, anxious to ascertain the efficiency of the system. The only doubt which seemed to have any justification had reference to the manœuvring of the vessel alongside the piers. There was some difficulty in this, due to lack of experience by navigating officers, who had formerly been employed on paddle steamers, and it ceased

Plate LIII.—Clyde Turbine Steamer "Queen Alexandra" (Fig. 159).
(Speed attained, 21.625 Knots.)

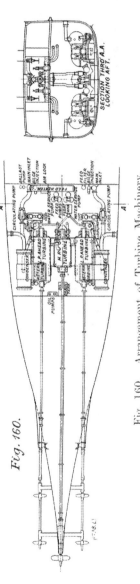

Fig. 160.

SECTION THRO'A.A.
LOOKING AFT.

Fig. 160. Arrangement of Turbine Machinery.

Fig. 161. The Dover and Calais Mail Steamer "The Queen"; Speed attained, 21.76 Knots.

Plate LIV.—First Cross=Channel Turbine Steamer.

when experience was gained with the turbine ships. There was also doubt as to the adequacy of the power for going astern, but this was not an inherent difficulty, it was only a question of the size of turbines necessary to meet the conditions of service.

The first of the English companies influenced by the performance of the two Clyde boats was the South-Eastern Railway Company, who, in 1902, placed an order with Messrs. Denny, of Dumbarton, for the building of the first turbine cross-Channel steamer, for service between Dover and Calais. This vessel—"The Queen"—was 310 ft. long, 40 ft. beam, and 25 ft. depth. On her trial trip on the Clyde the vessel steamed 21.76 knots ahead and 13 knots astern. Starting and stopping trials were also carried out, and when steaming 19 knots the vessel was brought to a dead stop in one minute seven seconds from the time that orders were given from the bridge, while the distance travelled was equal to only two-and-a-half times the length of the ship, or 770 ft. It was also observed that she gathered way from "rest" more quickly than paddle steamers. The arrangement of turbines was generally the same as in the case of the Clyde steamers, as will be seen by reference to the plan, Fig. 160 on Plate LIV., on which also is a view of the vessel under easy steam.

The working of this steamer attracted international interest, largely because of the service in which she was employed, and, therefore, some comparison may be made of the performance of this vessel and the paddle steamers in the same service. These results are set out in diagrammatic form on the next page (Fig. 162), alongside the results of three of the most efficient paddle steamers on the same service fitted with compound engines. The particulars given comprise the chief items of relative earning power and expenditure. The number of passengers which may be carried has been accepted as the basis of comparison. The shaded columns in each case represent the results of the turbine steamer; the open columns those of three paddle steamers. It will be seen that in columns marked No. 1 the coal consumed per passenger is from 25 to 107 per cent. more in the paddle steamers than in the turbine steamer. The engine room staff shown by No. 2 is much less. The oil consumption, indicated in No. 3, is enormously less. The cost of coal. oil, and engine room staff per passenger is from 33 to 106 per cent. less in the turbine boat than in the paddle steamers, as shown combined in No. 4. The turbine vessel, too, made a much larger number of runs, as shown in No. 5, not only because of her great popularity, but also because she was faster, taking nine minutes less on the twenty miles' run from Calais to Dover than the fastest boats, and because she was a better sea-boat.

It has been established beyond doubt that a turbine vessel can

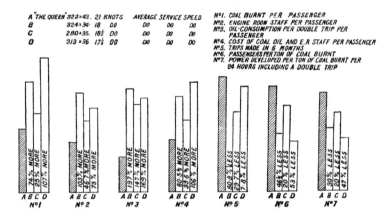

FIG. 162. DIAGRAM SHOWING RESULTS OF TURBINE-PROPELLED CROSS-CHANNEL STEAMER, "THE QUEEN," COMPARED WITH RESULTS OF THREE PADDLE STEAMERS FITTED WITH ORDINARY ENGINES, ON THE CALAIS-DOVER ROUTE.

more easily be driven in a heavy sea without danger to the machinery than any vessel fitted with other types of machinery. The smaller turbine-driven screw propellers do not emerge from the water when a heavy sea is running, as do the larger propellers in piston-engined boats; nor does rolling due to a beam sea affect them to the same degree as it does paddle-wheel steamers. In the remote contingency of the turbines racing, owing to these causes, no damage can result, as is the case with other types of engines. Thus the turbine vessel need never slacken speed to comply with considerations of safety of the machinery.

No. 6 column in Fig. 162 shows that the other ships take 20 to 53 per cent. less passengers per ton of coal consumed than

Plate LV. — Cross-Channel Steamer "Brighton" Steaming 21.37 Knots (Fig. 163).

Plate LVI.—Cross=Channel Turbine Steamer " Princess Maud " Steaming 20.66 Knots (Fig. 164).

does the turbine vessel. The final comparison (No. 7) is of power developed per ton of coal burnt per twenty-four hours, including a double trip, and here again the turbine comes out most satisfactorily.[1]

For the London, Brighton, and South Coast Railway Company Messrs. Denny, of Dumbarton, completed — in 1903 — a turbine steamer for the Newhaven and Dieppe service. This turbine vessel, the "Brighton," illustrated on Plate LV., facing page 102, assisted further in establishing the economy and popularity of the system, especially in rough weather. She was rather smaller than "The Queen," having a length of 280 ft., a beam of 34 ft., and a depth of 15 ft. $2\frac{1}{2}$ in. ; her turbines, driving three shafts, were of 6500 equivalent indicated horse-power, and the speed realised was 21.37 knots.[2]

The high-pressure turbine ran at 523 revolutions and the low-pressure at 577 revolutions per minute. These speeds were found quite satisfactory from the point of view of turbine design, as they enabled the turbines to be relatively small in diameter in proportion to the power required. Thus in the "Brighton" the diameter of the high-pressure turbine was $41\frac{1}{4}$ in. at the high-pressure end and 45 in. at the low-pressure end for five-fold expansion, while in the low-pressure turbines the diameters increased in steps from 42 in. to 54 in. with 25-fold expansion. It was possible, therefore, to give comparatively long blades, the length of blade in the high-pressure turbine being from $1\frac{1}{8}$ in. to 3 in., while in the low-pressure turbine the lengths were from $1\frac{1}{2}$ in. to $7\frac{1}{2}$ in. This led to small relative clearances, averaging about 2 per cent. of the length of the blade, and this contributed to efficiency.

There was, however, still room for improvement in this direction, which was achieved in later vessels by thinning the tips of the blades to chisel sharpness, so that, in the event of rubbing, the danger of damaging the blades was reduced to a minimum. In one or two cases where the adjustment had been tampered

[1] "Proceedings of the Institution of Civil Engineers," vol. clxiii. (Session 1905-6), Part I., page 183.

[2] See "ENGINEERING," vol. lxxvi., page 325.

with, the misfortune of stripping the blading arose, and resulted in the introduction of this system of thinning the tips of the blades as described in the previous chapter, page 49.

The next steamer built, also by Messrs. Denny, of Dumbarton, was the " Princess Maud," for the Larne and Stranraer service. This vessel is illustrated on Plate LVI., facing page 103. Comparisons were possible in this service also between the turbine and the existing paddle steamers, and the result was considerably to the advantage of the newer system.

Meanwhile opportunity had been taken in connection with the " King Edward " and " Queen Alexandra " for a thorough investigation of the propeller problem, different propellers being fitted to the vessels at various periods. The data collated, in association with the careful analysis of results in connection with warship trial performances, added very considerably to the knowledge of this difficult problem. A further step in the propeller problem was taken in connection with the " Emerald," a yacht launched in 1902 by Messrs. A. Stephen and Son, Limited, Scotstoun, Glasgow, and illustrated on Plate LVII., facing this page.

The " Emerald " is one of three turbine yachts completed in 1903. This vessel is 198 ft. long, 28½ ft. beam, and 18½ ft. deep, and when displacing 900 tons has a speed of 14½ knots. She was fitted originally with three turbines, corresponding in arrangement to the preceding ships, the high-pressure turbine driving the centre shaft with one propeller, while on each wing shaft, rotated by a low-pressure turbine or an astern turbine, there were two propellers.

The disadvantage of this arrangement of propellers was subsequently recognised. In view of the apparent inefficiency of the first set of propellers tried in this vessel, and the local vibration and noise in that part of the ship adjacent to the forward wing propeller on each side, it was decided to try the effect of entirely removing the forward of the two propellers on the wing shafts : the result was an increase of a quarter of a mile per hour to the speed of the ship for about the same power. When, however, the single propeller on each shaft was made larger, the speed for the same power was augmented to the extent of about half a mile

Fig. 165. Yacht "Emerald"; Speed attained, 15.0 Knots.

Plate LVII.—The First Turbine Steamer to Cross the Atlantic.

Fig. 166. Yacht "Tarantula"; Speed attained, 25.36 Knots.

Plate LVIII.—First Vessel with Separate Cruising Turbines.

per hour. The loss of efficiency observed appears to have been due partly to interference from the forward screws and partly to cavitation. Subsequently the "King Edward" and the "Queen Alexandra" were fitted with only one propeller on each shaft, and this practice has been adopted in all subsequent vessels. The "Emerald," too, was the first turbine vessel to cross the Atlantic Ocean. She is now owned by Lord Furness, and he has had the machinery altered, a reciprocating engine taking the place of the high-pressure turbine on the centre shaft, and exhausting into the low-pressure turbines on the wing shafts. The "Emerald" was the first yacht to have this system of piston-engine and exhaust-steam turbine—a system which is dealt with in a later chapter of this work.

In 1902 Messrs. Yarrow and Co., Limited, built the "Tarantula" for the late Colonel McCalmont. This vessel, illustrated on Plate LVIII., facing this page, is 152 ft. 6 in. in length, 15 ft. 3 in. in beam, and 8 ft. 9 in. in depth. She was fitted with a high, intermediate, and low-pressure turbine working in series on three shafts as in the "Turbinia." A cruising turbine was fitted at the forward end of the low-pressure turbine for use at speeds up to 15 knots. This was the first instance of cruising turbines, adopted later in several warships, as already described in the preceding chapter, and further considered in the later chapter on recent developments. The mean speed at maximum power on trial was 25.36 knots on a displacement of 150 tons. The turbines in this case were of 2000 shaft horse-power. This yacht was subsequently purchased by Mr. W. K. Vanderbilt, junr., of New York.

Another yacht, the "Lorena," was built in 1903 by Messrs. Ramage and Ferguson, Leith, for Mr. A. L. Barber. She has a length of 253 ft., a beam of 33 ft. 3 in., a depth of 20 ft. 4 in., and a displacement tonnage of 1700 tons. With three turbines and three screw propellers she obtained a speed of 18.02 knots.

In 1904 another important step in the establishment of the superiority of the turbine was made. The Midland Railway had four steamers built for their new service between Heysham and Belfast and the Isle of Man.[1] Professor J. Harvard Biles, of Glasgow University,

---

[1] See "ENGINEERING," vol. lxxviii., pages 423, 499.

who was responsible for the design of the vessels, recommended the adoption of the turbine system in two of the vessels. Three of the vessels, for the Belfast trade, were of the same model. The "Londonderry," which was built by Messrs. William Denny and Brothers, Dumbarton, was fitted by The Parsons Marine Steam Turbine Company with turbines similar to those in the Channel steamers previously referred to; while the "Antrim," built by Messrs. John Brown and Co., Limited, Clydebank, and the "Donegal," built by Messrs. Caird and Co., Limited, Greenock, had four-cylinder triple-expansion engines. The fourth steamer, the "Manxman," built by the Vickers Company, Barrow-in-Furness, and equipped with turbines by the Parsons Company, differed somewhat in her design, being intended especially for excursion traffic to the Isle of Man; she had more extensive passenger promenading area, with fewer cabins. The "Manxman" is illustrated on Plate LIX., facing this page. In respect of her machinery there was this further departure, that the boiler pressure was 200 lb. instead of 150 lb. per square inch, as in all preceding turbine vessels. While the other vessels are 330 ft. long, 42 ft. beam, and 18 ft. depth, moulded, or 25 ft. 3 in. from the promenade deck, the displacement being 2200 tons, the "Manxman" is 330 ft. in length and 43 in. beam, displacing 2330 tons.

The "Londonderry" has turbines of sufficient power to maintain the same speed as the vessels with reciprocating engines; and the weight of the machinery, including everything associated with the propelling of the ship, is, in the "Antrim" 730 tons, and in the "Londonderry" 575 tons. The weight of the boilers of the "Londonderry" is 15 per cent. less than in the case of those of the "Antrim," which weigh, with water, 460 tons. The main reciprocating engines weigh 210 tons as compared with 160 tons for the turbine machinery of equal power. The weight of shafting is 60 and 25 tons respectively. The total weight of the "Manxman's" machinery, notwithstanding the higher power developed, is 15 per cent. less than that of the "Antrim."

On Plate LXII., facing page 107, there are half-cross-sections through the piston engine room of the "Antrim," and a section

Plate LIX.– Cross=Channel Turbine Steamer "Manxman" Steaming 23.0 Knots (Fig. 167).

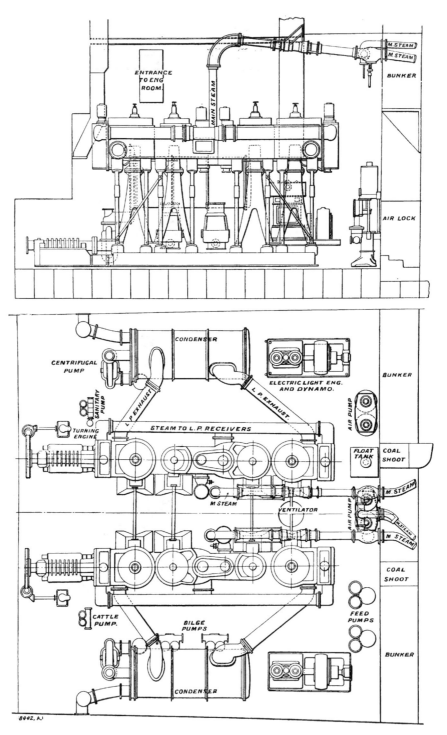

Plate LX.—Quadruple=Expansion Machinery of "Antrim"
(Figs. 168 and 169).

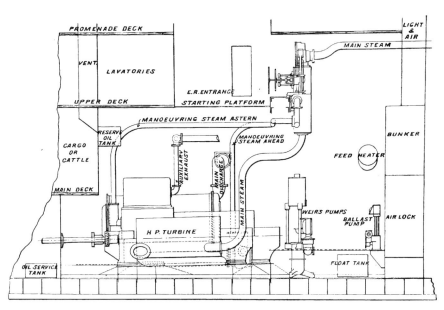

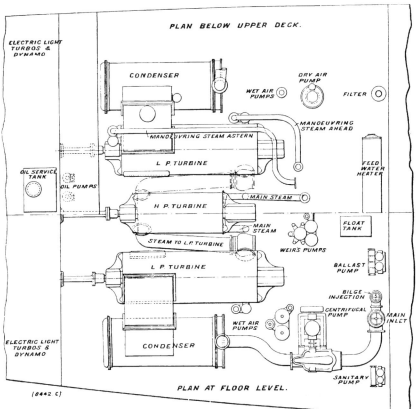

Plate LXI.—Turbine Machinery of "Londonderry"
(Figs. 170 and 171).

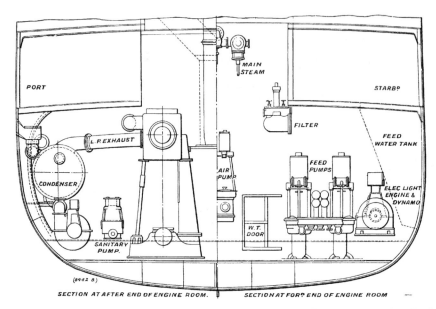

SECTION AT AFTER END OF ENGINE ROOM.   SECTION AT FORD END OF ENGINE ROOM

Fig. 172. Quadruple-Expansion Engines and Auxiliary Machinery in "Antrim."

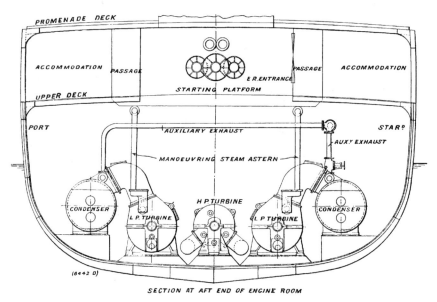

SECTION AT AFT END OF ENGINE ROOM

Fig. 173. Turbines in "Londonderry."

**Plate LXII.— Comparison of Piston Engines of "Antrim" and
Turbine Engines of "Londonderry."**

through the turbine engine room of the "Londonderry"; while on Plates LX. and LXI. there are similarly comparative longitudinal sections and plans of both sets of machinery. Although no great attempt was made to economise room in the turbine ship, it will be recognised that less space is actually required.

In the "Manxman" the designers decided to adopt the same steam pressure as in the piston-engined ships—namely, 200 lb.— the turbines being proportioned accordingly. Simultaneously there was fitted for the first time to a merchant ship Parsons vacuum augmenter, in which a jet of steam acting through a nozzle assists to withdraw the air from the main condenser, as described in a previous chapter (page 58). The "Manxman" had a speed $\frac{3}{4}$-mile per hour higher than the "Londonderry"—23 knots as compared with $22\frac{1}{4}$ knots for the "Londonderry," and 21.9 knots for the "Antrim." The revolutions on the centre shaft with the high-pressure turbine were 552 for the "Manxman" and 665 for the "Londonderry," and on the wing shafts, run by the low-pressure turbines, 618 revolutions for the "Manxman" and 755 for the "Londonderry." The higher speed of the "Manxman" represented a gain of 14 per cent. in power, and the efficiency of the turbine and propeller was at least 12 per cent. higher at 20 knots than that of the engine and propeller in the piston-engined ship.

The "Loongana," built by Messrs. Denny, of Dumbarton, and fitted with turbines by the Parsons Company, for the Union Steamship Company, of New Zealand, was the first turbine vessel to make the voyage to the Antipodes. This vessel was about the same size as the Channel steamers, being 300 ft. long, 43 ft. beam, and $16\frac{1}{2}$ ft. depth, with a displacement of about 2300 tons. On trial this vessel made a speed of 20.15 knots, the turbines being of 5000 horse-power. The voyage to New Zealand was made at a speed of 15 knots, the net steaming time being $30\frac{1}{2}$ days. The turbines worked well, and the maximum non-stop run was eight days.

The next turbine vessel to cross the Atlantic was the "Turbinia," built for service on Lake Ontario by Messrs. Hawthorn, Leslie, and Co., and engined by the Parsons Company. This was in 1904.

The vessel was 250 ft. long and 33 ft. beam, with the three-shaft turbine arrangement which had become practically universal for such vessels. The equivalent indicated horse-power was about 3500, and the speed attained was 18.46 knots.

The efficiency of the various Channel steamers, the reliability established by the run of the " Loongana " to New Zealand, and the success of the destroyers and cruisers built for the Navy— notably the " Eden " and " Amethyst "— as already established (page 82 *ante*), had removed all doubt as to the efficiency of the Parsons turbine, and from this time forward there was a steady development.

Messrs. Denny, of Dumbarton, continued, as a result of their experience, to be strong advocates for the system, and in 1904-5 they built four vessels for the British India Steam Navigation Company for service between the Persian Gulf and India—the " Lhasa," " Linga," " Lama," and " Lunka." These vessels are 275 ft. long, 44 ft. beam, 25½ ft. depth, and have a speed of 18 knots. A notable point in connection with these vessels is that the " Lama " and " Lunka " had their Parsons turbine installations manufactured by Messrs. Denny and Co., of Dumbarton. Up to this point every steamship fitted with turbines had had the engines from the Wallsend Works of The Parsons Marine Steam Turbine Company, but the system was now being so largely adopted that other engineering firms sought licences to construct the Parsons turbine from the designs of the inventor, and it was only appropriate that Messrs. Denny of Dumbarton should be the first to complete a vessel so fitted.

About the same time two large steamships for the Canadian Trans-Atlantic Service were ordered by the Allan Line, and fitted with turbines. This was a notable step in advance in the application of the new system to ocean-going ships of high speed, and the Allan Company—and particularly Sir Nathaniel Dunlop, the Chairman of the Company then—deserve great credit for being pioneers in the adoption of the system for large ocean ships. The two Atlantic liners for the Allan Line were the " Virginian," constructed by Messrs. A. Stephen and Sons, Limited, and the " Victorian," built

Fig. 174. The Allan Liner "Virginian" Steaming 19.11 Knots.

Plate LXIII.—The First Turbine Atlantic Liner.

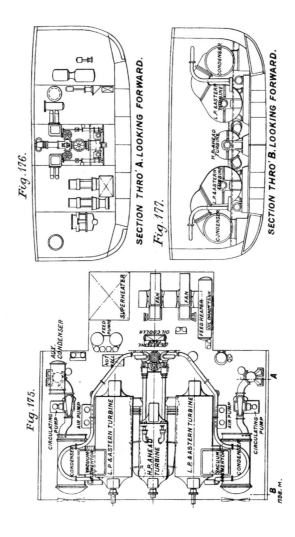

Fig. 176.

SECTION THRO' A. LOOKING FORWARD.

Fig. 177.

SECTION THRO' B. LOOKING FORWARD.

Fig. 175.

Plate LXIV.— Arrangement of Turbine Machinery in the "Virginian" (Figs. 175 to 177).

by Messrs. Workman, Clark, and Co., Belfast. These vessels were 520 ft. long, 60 ft. beam, and 41.2 ft. deep, displacing at 29 ft. 6 in. draught 13,000 tons. A view of the "Virginian" is given on Plate LXIII., facing page 108.

The turbines of the "Virginian" were made by the Parsons Company, and those for the "Victorian" by Messrs. Workman, Clark, and Co. They were considerably larger than any previously constructed. The high-pressure turbine had a diameter at the high-pressure end of $68\frac{3}{4}$ in., and at the low-pressure end of 74 in., while the low-pressure rotor was stepped in diameter from $95\frac{3}{4}$ in. to 107 in., to suit the range of expansion. The arrangement of the machinery is shown on Plate LXIV., facing this page.

In view of the dimensions of the casings it was found necessary to divide the castings into parts, to reduce their weight and size. The general design of bearings followed the practice adopted in the "King Edward," as illustrated, with the exception that the adjusting blocks were made stouter and heavier. Previous to the "Virginian," and including the "Victorian," the bearings were water-jacketed, but in the "Virginian" special oil coolers were introduced. Owing to the size of these turbines, it was not possible to obtain satisfactory lap-welded steel tubes for the rotors. In the case of the "Victorian" solid rolled forgings were fitted, and in the "Virginian" the drums were rolled from steel boiler plates with a butt strap-joint. The astern rotor drums were lap-welded with a cast-steel connecting ring to the low-pressure drum. The dummy drums for these turbines were made separate from both the drum and the wheel rim, as trouble had been experienced in obtaining satisfactory steel castings when the dummy formed part of the wheel rim. Steel forgings were adopted for the dummy ring, which ensured good metal being obtained for the turning of the dummy-ring grooves.

The revolutions at the full speed of 19 knots were only 290 per minute. It was computed that the weight of machinery was 400 tons less than if triple-expansion engines had been adopted, while the coal consumption on the service speed of 17 knots was practically the same as for reciprocating engines of the most modern type, being

1.4 lb. per equivalent indicated horse-power.   These vessels continue to give great satisfaction on service.

Another firm which took a licence at an early stage was the Fairfield Shipbuilding and Engineering Company, Limited, Govan, who, in 1904, fitted the new yacht "Narcissus" with turbines, but in this case only two shafts were adopted corresponding to the ordinary twin-screw system.   The high-pressure turbine operated the port propeller, and the low-pressure turbine the starboard screw, reversing turbines being incorporated with each.   This yacht, which was 210 ft. in length, 27 ft. 6 in. in beam, and 11 ft. 2½ in. in draught, had a displacement of 978 tons, and with 1600 effective horse-power, the speed was 14½ knots.   The results were satisfactory.

About the same time the turbine-propelled yacht "Albion," of 235 ft. in length and 1300 tons yacht measurement, was designed by Sir William White for the late Sir George Newnes, Bart.   She had triple shafts, the high-pressure turbine being placed on the centre shaft and the low-pressure turbines on the outer shafts.   On trial this vessel developed 2500 horse-power and attained a speed of over 15 knots.   The turbines were made at the Turbinia Works, while the vessel and her boilers were constructed by Messrs. Swan, Hunter, and Wigham Richardson.   This yacht has been continuously at work, has made numerous long voyages, and has proved both handy and economical in fuel consumption at the ordinary cruising speed of 12½ to 13 knots.

The year 1905 may be said to have been the turning point in the education of engineers as to the efficiency of the turbine.   Data continued to accumulate, but opinion had not completely crystallised in favour of the system.   The decision of the Cunard Company, late in 1904, to adopt the system in the "Mauretania" and "Lusitania," and the intention subsequently declared by the Admiralty to fit turbines to all warships, had a considerable effect on public opinion, especially as both steps were taken as the result of great deliberation.   When, in 1904, an agreement was entered into between the Government and the Cunard Steamship Company for the building of two vessels with an average sea-speed in moderate weather of not less than 24½ knots, to be utilisable

Plate LXV. The Turbine Cunard Liner "Carmania" (Fig. 178).

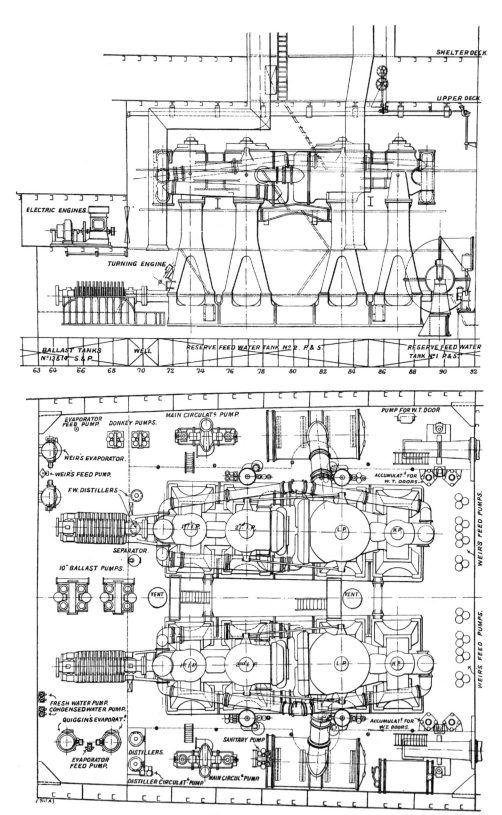

SHELTER DECK.

UPPER DECK.

ELECTRIC ENGINES.

TURNING ENGINE

BALLAST TANKS
Nº 13 & 14 S. & P.

WELL

RESERVE FEED WATER TANK Nº 2. P. & S.

RESERVE FEED WATER
TANK Nº 1 P. & S.

63 64   66   68   70   72   74   76   78   80   82   84   86   88   90   92

EVAPORATOR
FEED PUMP.

DONKEY PUMPS.

MAIN CIRCULAT'S PUMP.

PUMP FOR W.T. DOOR.

WEIR'S EVAPORATOR.

WEIR'S FEED PUMP.

F.W. DISTILLERS.

SEPARATOR.

10" BALLAST PUMPS.

VENT

ACCUMULAT'S FOR
W.T. DOORS.

WEIR'S FEED PUMPS.

1ST L.P.   2ND L.P.   L.P.   H.P.

VENT

H.P.   2ND L.P.   L.P.   H.P.

WEIR'S FEED PUMPS.

FRESH WATER PUMP.
CONDENSED WATER PUMP.
QUIGGINS EVAPORAT'S

EVAPORATOR
FEED PUMP.

DISTILLERS.

DISTILLER CIRCULAT'S PUMP.

SANITARY PUMP.

MAIN CIRCUL'S PUMP.

ACCUMULAT'S FOR
W.T. DOORS.

Plate LXVI.—Reciprocating Engines in the Cunard Liner "Caronia"
(Figs. 179 and 180).

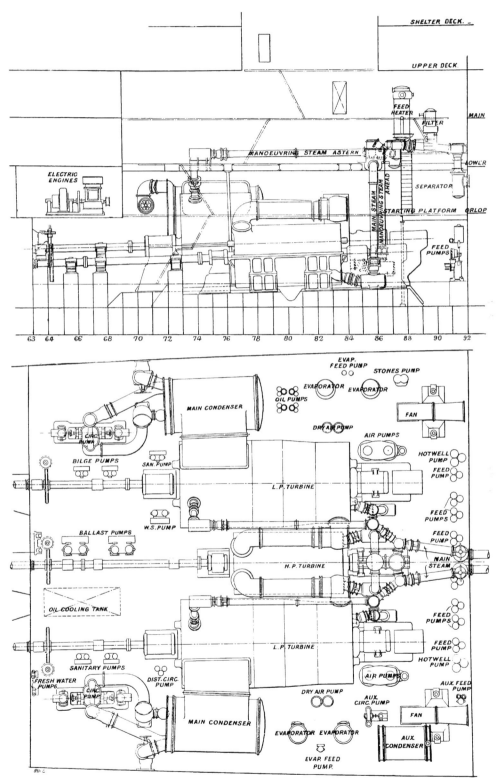

Plate LXVII.—The Turbine Engines in the Cunard Liner "Carmania" (Figs. 181 and 182).

Plate LXVIII.—The Turbines of the "Carmania" in Course of Construction in the Clydebank Works (Fig. 183).

not only for mail service, but for cruiser service in time of war; the problem of ensuring this speed greatly exercised several of the ocean steamship builders of the country. The earlier designs anticipated the adoption of reciprocating engines driving three shafts, but the result of many consultations between the prospective builders and the Admiralty, and the disclosure of the data attained not only in the tank at Haslar but in the turbine-driven steamships then in service, induced the late Lord Inverclyde, then the Chairman of the Cunard Company, to appoint a Commission to report upon the suitability of the Parsons turbine for the two new ships. This Commission included Mr. James Bain, then the Marine Superintendent of the Cunard Company; Engineer Vice-Admiral Sir H. J. Oram, K.C.B., now the Engineer-in-Chief of the Navy; Mr. J. T. Milton, Chief Engineer Surveyor of Lloyds' Registry; and the late Mr. H. J. Brock, of the firm of Messrs. Denny, of Dumbarton; while the three firms concerned in the building of the two new ships were represented—Messrs. John Brown and Co., Limited, by Mr. Thomas Bell, now Director of the Clydebank Works; Messrs. Swan, Hunter, and Wigham Richardson, by Sir William White, K.C.B.; and the Wallsend Slipway and Engineering Company by Mr. Andrew Laing, the Managing Director. This Committee most carefully considered the subject, and made water consumption trials on board merchant and naval ships as well as in land stations where turbines and reciprocating engines were working side by side. The result of their report to the Cunard Company was the decision to fit Parsons turbines to both vessels.

Before the Commission had finally reported, the Cunard Company decided to fit Parsons turbines to one of two vessels of 30,000 tons displacement, ordered in 1904 to be built by Messrs. John Brown and Co. The turbine vessel was named the "Carmania," and the piston-engine ship the "Caronia." The vessels, of the intermediate type, have a length of 678 ft., a beam of 72 ft., and a displacement of 30,918 tons at 33 ft. 3¾ in. draught. The engraving on Plate LXV. shows the "Carmania" on trial. The vessels were alike in design but the stern of the turbine-propelled ship had to be modified to suit the triple screw arrangement. The boiler installation was the same in both vessels, no advantage being taken of the reduction which has since become general in view of the lower steam consumption of the turbines for a given power. The general result was that, on trial, the "Carmania" proved herself capable of maintaining a speed of 20 knots, as compared with $19\frac{1}{2}$ knots in the case of the "Caronia," and that, too, for practically the same steam consumption. A comparison of the machinery arrangement of the two vessels is afforded by the drawings reproduced on

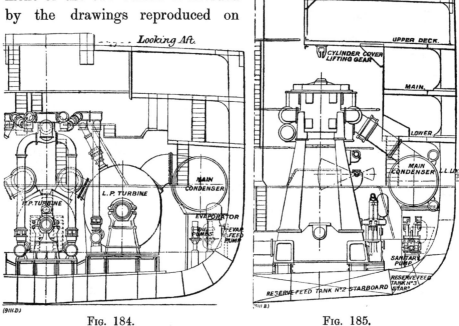

<table>
<tr><td>FIG. 184.</td><td>FIG. 185.</td></tr>
<tr><td>TURBINE ENGINES IN THE "CARMANIA."</td><td>RECIPROCATING ENGINES IN THE "CARONIA."</td></tr>
</table>

Plates LXVI. and LXVII., adjoining pages 110 and 111, while the cross sections, Figs. 184 and 185, on this page, show further the saving in space resulting from the adoption of the turbine.[1]

The turbines of the "Carmania" were made by Messrs. John Brown and Co., Limited, who had acquired a license from the Parsons Company, and a view of the three turbines in course of erection is shown on Plate LXVIII. This view is specially interesting as it

---

[1] See "ENGINEERING," vol. lxxx., page 715.

shows the method of lifting the top half of the casing and the rotor of the turbine. There are rods, as shown in the case of the high-pressure turbine in the centre, and in the low-pressure turbine to the right, of the engraving, which guide the cover, and a cast-iron crosshead used for raising the top half of the casing or the rotor. The crossheads are lifted by means of square-threaded screws and worm-gear, driven in the case of large-powered engines by electric motors, so arranged that both ends of the casing or rotor are raised simultaneously.

The results of the trials of the "Virginian," "Victorian," and "Carmania" not only justified the recommendation of the Cunard Commission, but assisted greatly towards developing public opinion in favour of the system, and in 1905-6 there were many applications of the system, not only for high-speed Channel steamers, but for vessels for service further afield. Messrs. John Brown and Co., Limited, fitted in the Clyde pleasure steamer "Atlanta" an experimental three-shaft installation, erected in their works to obtain some corroborative data respecting steam consumption.

The South-Eastern and Chatham Railway, as a result of their experience with "The Queen," added the "Onward" and "Invicta," both built by Messrs. Denny, of Dumbarton. These vessels showed a steam economy of 15 per cent. as compared with the preceding paddle steamers. The following refers to the cost of wear and tear in connection with these vessels :—

"The tonnage of the turbine steamer—one of Denny's build—is 27 tons greater, and the number of passengers carried 64 per cent. higher, than in the case of the steamer with compound diagonal engines, while the speed is 22 knots against 17 knots. The labour cost of making repairs in a season was for the turbines £176, and for the main engines of the paddle steamer £273, including, in both cases, the cost of opening out the machinery for the Board of Trade Surveys. These sums do not include the cost of material in respect of which the result is equally favourable to the turbines. The only renewal in three years in the turbines was one set of glands, costing about £90, whereas in paddle steamers renewal of piston packing rings, brasses or bushes, or some such detail, has been required every quarter. It is true that the paddle steamers are older by four or five years ; but the comparison of labour cost is made for a year, during which the turbine steamer made 520 trips against a paddle steamer which made 288 trips. Thus the turbine has clearly a most distinct advantage over ordinary engines in respect to wear and tear. In the same period the coal consumption was 24 per cent. in favour of the turbine vessel."

Q

The approved success of the turbine was further shown in 1907 when, for the same Company, Messrs. Denny built the "Victoria" and "Empress," so that now the South-Eastern and Chatham Railway have five turbine steamers, all confirming the high success achieved by "The Queen," the prototype of the fleet.

Another notable vessel was the "Viking," which was built in 1905 for the Isle of Man Steam Packet Company, of Douglas, by Sir W. G. Armstrong, Whitworth, and Co. This vessel, which is illustrated on Plate LXIX., facing this page, is 361 ft. in length overall, and 350 ft. between perpendiculars. The breadth is 42 ft. and the depth to the upper deck 17 ft. 3 in. One desideratum in the design was to secure the maximum accommodation for passengers, especially promenading space, as the vessel is engaged in excursion traffic between Liverpool and the Isle of Man. It was found that the adoption of the turbine, owing to the small headroom required by the turbine machinery, greatly facilitated the attainment of this aim. The turbines follow the design of immediately preceding vessels, there being three shafts with the high-pressure turbine on the centre and the low-pressure turbines on the wing shafts, the astern turbines being incorporated within the same casings as the latter.

The mean speed realised on two runs, one in each direction, over a 68¼-mile course in the North Sea was 23.53 knots. Although the service speed is 22.2 knots, the coal consumption per nautical mile steamed is only 0.472 ton as compared with 0.614 ton in the case of a vessel of practically the same dimensions but only of 20-knot speed. This latter vessel has three-cylinder compound paddle engines; owing to their greater weight the ship's displacement is 540 tons more. The number of passengers carried is practically the same in each ship. The engineer's staff is less in the "Viking" by one engineer and two greasers, so that there is an appreciable difference in the coal, oil, and wages bill.

In 1905, also, as a consequence of the success of the "Brighton," there was ordered by the London, Brighton, and South Coast Railway Company, for their service between Newhaven and Dieppe, the "Dieppe," which was built and engined by the Fairfield Shipbuilding and Engineering Company. As with the "Brighton,"

Plate LXIX.—The Isle of Man Steamer "Viking"; Speed attained on Six Hours' Trial, 23.53 Knots (Fig. 186).

Plate LXX.—The Newhaven and Dieppe Mail Steamer "Dieppe" Steaming 21.64 Knots (Fig. 187).

Photo by J. H. Gear, London.

severe limitations were imposed in connection with the design of the vessel by the dimensions and draught of water in harbours, so that she had a length of only 284 ft., a beam of 34 ft. 8 in., and a depth of 22 ft. 1 in. Notwithstanding these limitations to dimensions, the vessel on the final trial, when carrying a specified load, had a draught of only 9 ft. $5\frac{3}{4}$ in., and attained a mean speed on the run to and from Dieppe of 21.64 knots, about half a mile per hour in excess of that required by the contract. The attainment of these results was largely a consequence of the adoption of the turbine. In the running of this vessel, as compared with steamers driven by piston engines of corresponding size and power, experience has shown that the turbine vessel is the better sea boat, and is less influenced by severe weather conditions, particularly in respect of speed. The turbine steamer, with the same power and coal consumption, gains three or four minutes in fine weather on the three hours' run between Newhaven and Dieppe, and about five minutes in rough weather, on vessels having reciprocating engines. In the " Dieppe " the astern turbines were of somewhat greater efficiency than in previous vessels, and on her official trials she was stopped in a distance equal to one and a half times her length, when steaming at 12 knots, the time taken to bring her to rest being thirty-one seconds. The " Dieppe " is illustrated on Plate LXX., facing this page.

Reference is justified to the " Maheno," built and engined by Messrs. Denny, of Dumbarton, also in 1905, for the Union Steamship Company of New Zealand, as she was the second turbine vessel to go to New Zealand. In this case the length of the vessel was 400 ft., the beam 50 ft., and the depth $30\frac{3}{4}$ ft. Her speed was $16\frac{1}{2}$ knots, and on the trip to New Zealand, where the vessel has since conducted a passenger service around the coast, the turbines worked admirably.

In 1906 the same firm added to the British India Steam Navigation Company's fleet a fifth turbine steamer, the " Rewa," of 455 ft. in length, about 10,000 tons displacement, and 19 knots speed; while in 1907 there was added to the Union Steamship Company's fleet in New Zealand, also by Messrs. Denny, the " Maori," a vessel 350 ft. in length, 47 ft. beam, and 3150 tons displacement, which on her trials attained a speed of 18 knots. In the same year this firm sent the first turbine steamers to Japan—the "Hirafu Maru" and the "Tamura Maru."

All the British railways have since recognised the advantage of the turbine system, and consequently there is no passenger service from the British Isles to Ireland or the Continent without one or more turbine steamers.

The Great Western Company inaugurated their Fishguard and Rosslare Steamship service, with three turbine steamers ordered in 1906, the "St. George" and the "St. Patrick," built by Messrs. John Brown and Co., Limited, of Clydebank, and the "St. David," built by Messrs. Cammell Laird and Co.[1]; while later they had a fourth vessel, the "St. Andrew," built by the first-named firm. An interesting departure was made in the turbines of these vessels.

DUMMY STRIPS

FIG. 188.   FIG. 189.

Difficulty having been experienced with the contact dummy in the turbines of one of the previous vessels, the rotor dummies having been allowed to come in contact while running on service, the method of supporting the rotor dummy ring was modified. The flange of the loose dummy, already referred to, was placed in such a position that in the case of an accidental contact of the dummies of the casing and rotor, the heat generated would cause the rotor drum to expand in the same direction as the casing dummy and thereby obviate any possible damage to the dummies. The dummy strips, which had hitherto been made of the section shown by Fig. 188, were shaped with fine tips, as in Fig. 189; these grind away and clear themselves in the event of accidental contact. The cylinder dummies were first made separate from the main casing casting in this vessel. The engraving on Plate LXXI., facing this page, shows the "St. George" steaming over 23 knots on trial, while on Plate LXXII., facing page 117, there is a view across the engine-room showing clearly the governor gear on the forward end of the turbines, and, to the right, the starting platform and the

[1] See "ENGINEERING," vol. lxxxii., page 106.

Plate LXXI.   The Great Western Railway Company's Fishguard and Rosslare Steamer "St. George" Steaming over 23 Knots (Fig. 190).

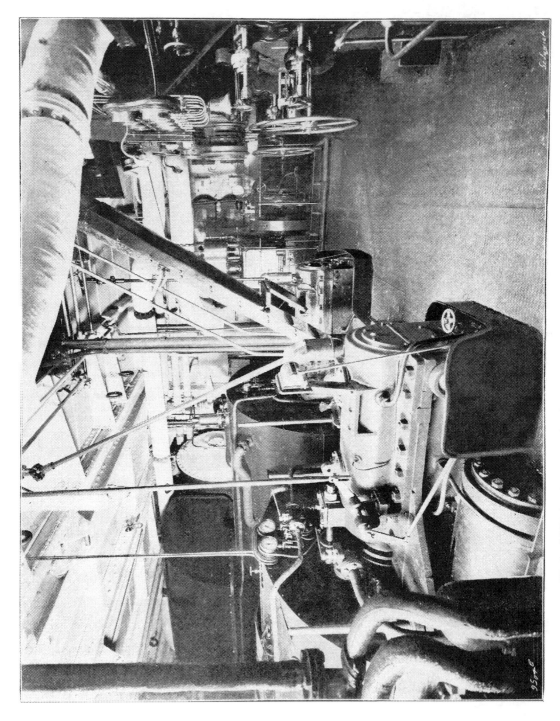

Plate LXXII.—View in Engine Room of the "St. George" (Fig. 191).

wheels for operating the steam valves. These are simple in arrangement and effective in control. In addition to the main steam valve, there is conveniently placed a three-way valve by the working of which steam is passed to the ahead or astern turbine or cut off entirely, so that a turbine steamer is as quickly manœuvred as is possible with mechanism.

The Great Eastern Railway Company, in 1907, ordered from Messrs. John Brown and Co. the " Copenhagen," and since then have put upon the service two duplicate steamers.

In 1908 the Lancashire and Yorkshire and the London and North Western combined service was augmented by two turbine steamers, built by Messrs. Denny,[1] while in 1909 the London and South Western Railway Company instructed Messrs. Cammell Laird and Co. to supply two vessels for their Channel Island traffic.

We do not propose, however, to attempt any complete account of the turbine work of recent years, but one or two outstanding cases should be put on record before we come to refer to the most important of turbine installations — those of the " Mauretania " and " Lusitania." It is worth noting that the first turbine steamer for the Thames traffic was the " Kingfisher," built in 1906 for the General Steam Navigation Company by Messrs. Denny, of Dumbarton ; while the first Glasgow and Belfast turbine steamer was the " Viper," built in 1906 by the Fairfield Company for Messrs. G. and J. Burns, Limited.

The " Ben-my-Chree " is notable because of the exceptional speed attained on trial, and the efficiency she has demonstrated in service. This vessel[2] was built by the Vickers Company at Barrow-in-Furness. She has a length of 375 ft., a beam of 46 ft., and a depth of 18 ft. 6 in. The arrangement of turbines conforms with almost universal practice. The starboard and central propellers are right-handed and the port left-handed. The diameter of the rotor drums are :—High-pressure, 3 ft. 11 in. ; low-pressure, 5 ft. 7 in. ; and the astern, 4 ft. 2 in. On the six hours' trial the " Ben-my-Chree " maintained an average speed of $24\frac{1}{2}$ knots with ease, and

---

[1] See "ENGINEERING," vol. lxxxviii., page 757.

[2] See "ENGINEERING," vol. lxxxvi., page 203.

for a portion of the run her speed was $25\frac{1}{2}$ knots. The astern speed was 16.6 knots on the measured mile, and when going 23 knots ahead she was brought to rest in a distance equal to three of her own lengths. The mean speed for twelve consecutive runs in deep water between Liverpool Bar and Douglas Head proved to be 24.12 knots; the mean draught then was 13 ft. 5 in., and the displacement 3353 tons. The coal consumption on these runs worked out to 0.53 tons per mile steamed, notwithstanding the very high speed. This compares with 0.47 tons for the 22.2 knots of the "Viking," and with 0.614 for the 20 knots speed of the immediately preceding paddle steamer of corresponding size and of less passenger capacity. The "Ben-My-Chree" has a passenger certificate for 2500 passengers. Her smart appearance can be appreciated from the engraving on Plate LXXIII.

Reverting to sea-going passenger steamers of large size and high speed, allusion must be made to two notable ships, the "Cairo" and "Heliopolis,"[1] built by the Fairfield Company in 1907 and originally designed for the Mediterranean service between Alexandria and Marseilles. The ships are 545 ft. in length, 11,000 tons gross register, and accommodate 709 first-class and 281 second-class passengers. These vessels—also fitted with triple-screw turbine engines—attained a speed of over $20\frac{1}{2}$ knots on a twelve hours' trial at 21 ft. 5 in. draught, the shaft horse-power being about 19,000. The steam consumption for all purposes worked out at about 14.94 lb. per horse-power per hour. Regarding their design, the late Dr. Elgar, in his Forrest Lecture to the Institution of Civil Engineers,[2] said that had reciprocating engines been fitted the boiler power would have required to be 6 per cent. greater, and the weight of the engines, boilers, and auxiliaries 400 tons greater. They were employed only for a short time in the Mediterranean; their steaming performances had been entirely successful, but they were withdrawn from the special service entirely for commercial reasons. The vessels have since been acquired by the Canadian Northern Railway Company and are now running

[1] See "ENGINEERING," vol. lxxxv., pages 560 and 616.
[2] See "ENGINEERING," vol. lxxxiii., page 825.

Plate LXXIII.—The Isle of Man Turbine Steamer "Ben=my=Chree"; the Fastest Channel Steamer Afloat. Maximum Speed, $25\frac{1}{2}$ Knots (Fig. 192).

Plate LXXIV.—The Atlantic Liner "Royal Edward," formerly the Egyptian Liner "Cairo," (Fig. 193).

between Bristol and Canadian ports. Their passenger accommodation has been modified and their fuel supply increased; and in their earliest voyages they created a record for the voyage to Canada. The vessels are now known as the "Royal Edward" and "Royal George." A view of the former, as refitted for the Atlantic service, is given on Plate LXXIV.

Practically from every point of view the most remarkable ships driven by turbines are the "Mauretania" and "Lusitania," The consideration given to the question of the type of machinery to be adopted has already been referred to on page 111. In connection with both ships much research work was undertaken, not only to determine the relative efficiency of three or four shafts, but also to arrive at satisfactory details in connection with the turbines. The record of experiments, as well as the complete details of design and method of construction, have been the subject of special numbers of "ENGINEERING,"[1] so that we need only deal here with those points which help to elucidate the principles of design and to establish the efficiency of the system.

Four shafts were adopted, partly because it was not considered prudent to transmit more than 20,000 horse-power through one shaft and one propeller. The arrangement also conferred the advantage that the power was equally divided on each side of the centre line of the ship, and thus assisted towards the manœuvring capability of the vessel. It enabled the wing propellers to be removed a reasonable distance from the skin of the ship, whereby vibration of the structure of the ship was reduced.

The general arrangement of the turbine machinery is illustrated by the drawings on Plate LXXV., facing page 120. There are six turbines in all, two high-pressure and two low-pressure turbines for going ahead, and two high-pressure units for going astern. The two high-pressure ahead turbines are placed in the wings and drive the outer shafts, while the two low-pressure ahead turbines drive the two inner shafts. Forward of these latter are located the high-pressure astern turbines. The two condensers are each in a separate compartment abaft the low-pressure ahead turbines, and the circulating

---

[1] See "ENGINEERING," vol. lxxxiv., page 129 ; *Ibid.* page 609.

and air pumps are in another compartment still further aft. The
hot feed-well, evaporators, feed-water heaters (on an elevated platform), &c., are at the forward end of the main engine rooms, where
also are the boiler-feed pumps. The auxiliary engine condensers, &c.,
are in the wings, abaft the high-pressure turbines.

The low-pressure shafts are at 9 ft. 6 in. centres from the middle
line of the ship, while the high-pressure shafts are 27 ft. from the
middle line, so that the distance between these two shafts is 17 ft. 6 in.
The shafts are parallel with the middle line, but have a slight angle
to the level of the keel, the rate of elevation in the case of the low-
pressure turbine being about $\frac{2}{10}$ in. per foot, and in the high-pressure
turbine rather less than $\frac{1}{2}$ in. per foot. This practice has been
adopted in almost all turbine vessels, the aim being to ensure
efficient drainage of the turbines and greater immersion of the screw
propellers, which reduces the possibilities of the turbine racing
in the event of the ship pitching in a following sea. The high-
pressure turbines are in advance of the low-pressure turbines, the
centre of the former being about 20 ft. ahead of the centre of the
latter, while the astern turbines are still further forward, the inter-
vening space being about 10 ft., and within this there is a bearing
and the thrust blocks.

There was an important difference in connection with the material
for the turbines for these vessels. In the one case the practice up
to this stage of adopting cast steel discs for the rotor drums was
followed : in the other instance forgings were used, a procedure which
is now general in all large installations as it makes for reliability.

The high-pressure turbine rotor is 96 in. in diameter, the low-
pressure 140 in., and the astern 104 in. One of the "Lusitania's"
rotors is illustrated on Plate LXXVI., while the low-pressure
turbine of the "Mauretania" is shown in Plate LXXVII., the
upper part of the casing being raised by means of the crosshead and
guide rods as already described in connection with the "Carmania."
The exhaust port is seen at the left hand on the casing top. The
blades in the high-pressure turbine range from $2\frac{1}{4}$ in. in length to
$12\frac{3}{8}$ in., and are arranged in eight groups of 15 rows on the rotor.
In the case of the low-pressure turbines the length of the blades

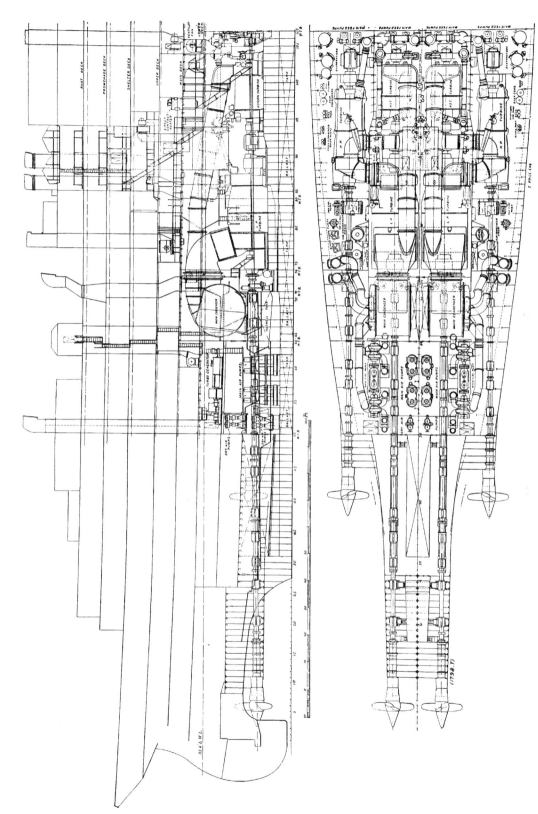

Plate LXXV.—General Arrangement of Turbine Machinery in the "Mauretania" (Figs. 194 and 195).

Plate LXXVI.—One of the "Lusitania's" Rotors Completed in the Clydebank Engineering Works (Fig. 196).

Plate LXXVII.—Low=Pressure Turbine of the " Mauretania," with Top Half of Turbine Casing Raised (Fig. 197.)

Plate LXXVIII.—Complete Ring of Low=Pressure Rotor Blading for the " Mauretania " (Fig. 198).

ranges from $8\frac{1}{4}$ in. to 22 in., and in the astern turbine from $2\frac{1}{4}$ in. to $8\frac{3}{4}$ in. The view on Plate LXXVIII., facing this page, shows a complete ring of the largest blades of the "Mauretania." The rotor of each high-pressure turbine, complete, weighs 86 tons, of the low-pressure turbine 120 tons, and of the astern turbine 62 tons. The greatest diameter of the high-pressure shaft is 3 ft. and of the low-pressure rotor 4 ft. 4 in., these having central holes 21 in. and 34 in. in diameter respectively. The over-all length of the turbines, as measured by the shafting, including the bearing, is 45 ft. 8 in. in the case of the high-pressure, 48 ft. $1\frac{7}{8}$ in. in the case of the low-pressure, and 30 ft. $1\frac{1}{4}$ in. in the case of the astern turbines.

The arrangements made for steam distribution are very extensive. The steam passing through the main valve enters the separator of the spiral tube form, from which the water is drained off through a trap, or straight into the hot well. From this separator the steam passes through the high-pressure regulating valve (of the equilibrium type) or to a manœuvring shut-off valve of the piston type, and thence to the manœuvring valve, which is also of the equilibrium type. This last valve is only used for manœuvring, and the high-pressure turbine is then usually thrown completely out of action. Thus in going out of harbour or entering port, where there are frequent changes in the direction of rotation of the propellers, the two inboard shafts only are used, the manœuvring valve controlling all steam distribution. In ocean steaming, however, this manœuvring valve is not in operation, steam from the separator passing to the high-pressure regulating valves into the high-pressure turbines, thence to the low-pressure ahead turbines, and finally to the condensers. It is scarcely necessary to say that great attention was devoted not only to the design of these valves, but to the mechanism for manipulating them, and as a consequence the shafts are manœuvred as easily when going in or out of harbour as in any vessel afloat. Arrangements are made, with the aid of sluice valves, for passing the exhaust from the high-pressure turbine direct to the condenser. The eduction pipe from the low-pressure ahead turbine is seen in the elevation Plate LXXV. in the form of a curve; that from the astern turbine is on a gradient to clear the ahead turbine.

An interesting paper,[1] read by Mr. Thomas Bell, of Messrs. John Brown and Co., Limited, gave data of the trial performances and of early voyages.   The full-power trial consisted of a forty-eight hours' run between the Corsewell Light on the Firth of Clyde and the Longship Light in the South of England, the distance being something like 300 miles.   The "Lusitania" on the four runs over this course attained a speed of 25.4 knots, the mean shaft horse-power being 68,850, while the slip of the propeller was found to be 15 per cent.   The "Mauretania," on the other hand,

attained on a corresponding trial a speed of 26.04 knots with a slightly higher mean power. Annexed is a curve showing the power for progressive speeds, the highest rate attained on service throughout the complete Transatlantic run being 26.06 knots.

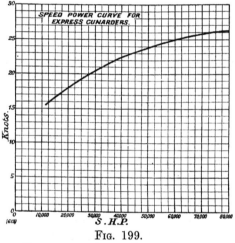

FIG. 199.

PROGRESSIVE SPEED CURVE, EXPRESS CUNARDERS.

Mr. Bell, in his paper already noted, showed that, on an early voyage, the average amount of coal burnt per hour for all purposes was 43½ tons.   The water evaporated per pound of coal was 10.2 lb., from a feed temperature of about 196 deg., equal to 10.9 lb. from and at 212 deg.   The coal for all purposes per shaft horse-power per hour was 1.5 lb., equivalent to 1.4 lb. per indicated horse-power per hour.[2]   Taking the mean displacement of 36,000 tons the figure given represents, at the speed named, a consumption of almost exactly 11 lb. of coal per 100 nautical miles per ton of displacement, a result which is very satisfactory, especially when it is

[1] "Transactions of the Institution of Naval Architects," vol. l., page 96 ; see also "ENGINEERING," vol. lxxxv., page 489.

[2] A German authority published, in the German technical journal *Schiffbau*, data which gave the comparative figures for one of the fastest German liners, fitted with piston engines, as 1.54 lb. per indicated horse-power per hour.

Plate LXXIX.—The Royal Yacht "Alexandra" (Fig. 200).

Plate LXXX.—The Turbines of the Royal Yacht "Alexandra" (Fig. 201).

remembered that only half of the coal was Welsh anthracite, the other half being Yorkshire. The steam consumption of the main turbines was 13.1 lb. per shaft horse-power per hour; but since then the economy has greatly improved, the steam consumption being reduced to about $11\frac{1}{2}$ lb. per shaft horse-power per hour.

The propellers have been changed in both ships consequent upon the experience gained and on further experiments, and the result has been still higher efficiency. The two ships have since been crossing the Atlantic in sunshine and storm at an average speed of between $25\frac{1}{2}$ and 26 knots. Almost each successive voyage established a new record, and the mean for the trip between Liverpool and New York has now exceeded 26 knots, a performance which, above all others, establishes the success of the turbine system. The highest speed by preceding vessels was 23.58 knots. It will thus be seen that on service the vessels have quite come up to the splendid performance of the trial trips. In this connection also it should be recalled that one of the cruisers of the Invincible class, the " Indomitable," fitted with turbines of 43,000 shaft horse-power, crossed from Canada in 1908 at an average speed of 24.8 knots, a record for long-distance voyages for warships. For 1684 nautical miles the mean speed was 25.1 knots. There is thus striking proof of the accuracy of the opinion expressed by Mr. Parsons ten years earlier as to the suitability of the turbine for Atlantic liners—an opinion quoted in the earlier paragraphs of this chapter.

High recognition was accorded to the efficiency of the turbine by its adoption in the new Royal Yacht " Alexandra," completed by Messrs. J. and A. Inglis, Limited, Glasgow, in 1908. In this vessel moderate draught was essential in order to enable her to enter many secondary ports, and the turbine machinery, with the smaller but faster running propellers, was particularly conducive to the reduction of draught without affecting propulsive efficiency. The new yacht is 275 ft. long, 40 ft. beam, and, on a mean load draught of 12 ft. 6 in., displaces 2050 tons; as shown on the engraving on Plate LXXIX., she has a very smart appearance.

The turbine machinery, which is illustrated on Plate LXXX., facing this page, as it stood in the erecting shop of The Parsons

Marine Steam Turbine Company, Limited, where it was constructed, is of the usual triple-shaft arrangement with the high-pressure turbine in the centre and the low-pressure and astern turbines on the wing shafts. This arrangement, as has already been pointed out, has great advantages in respect to manœuvring, as, in such case, the high-pressure turbine is disconnected, steam at full boiler pressure passing through a separate manœuvring valve to the low-pressure ahead or to the astern turbines. There is obtainable a very quick response, and, as the astern turbines are of high power, it is possible to reverse the motion of the ship within a very short period. In this respect the trials of the "Alexandra" proved exceptionally satisfactory. The arrangement of the governors is well shown on the engraving. The usual ball governor, it will be seen, is driven from the forward end of each shaft, and through a simple lever controls the steam valve of each turbine. The condensers are in the wings of the ship, and are fitted with the Parsons vacuum augmenter. The steam for the turbines is supplied by three large Yarrow water-tube boilers, which have, on exhaustive trials, proved quite satisfactory. On the official trials, at what is regarded in the Navy as "continuous steaming" speed, two boilers were sufficient to give $17\frac{1}{2}$ knots, while with all the boilers in use the speed was 19.15 knots, or nearly half a mile per hour in excess of the designed speed. Since completion the yacht has been on several cruises, and has given complete satisfaction.

These latest triumphs of the Parsons turbine, achieved within fifteen years of the first application of the system for steamships, attest indisputably not only the great mechanical success of the dominant idea of the invention, but the continuous exercise of ingenuity to meet the problems which beset the adaptation of the system to meet special conditions, speeds impracticable with piston engines have been realised with reduction in weight, fuel, first cost, and working expenses, and the reliability and simplicity of the machinery removes all causes of anxiety on the part of the staff.

THE TURBINE IN FOREIGN AND COLONIAL FLEETS.

AT the same time as the potentialities of the steam turbine for ship propulsion were being recognised in Great Britain, considerable interest was evinced by foreign marine constructors; but a noticeably larger degree of scepticism prevailed and prevented the development abroad from being as rapid as in Britain.

FRANCE.

Although the manufacture of land turbines began to be general on the Continent in 1900, it was not until 1902 that the first turbine-propelled vessel was ordered; and the turbines themselves were in this case imported from The Parsons Marine Steam Turbine Company's works at Wallsend. The vessel in question was the small French torpedo boat, No. 293, which remains to this day the smallest turbine vessel afloat, with the exception of the experimental "Turbinia." This boat, which is illustrated on Plate LXXXI., facing page 126, is remarkable for another reason—the cruising turbines which were applied until recently to all warships were first fitted to this vessel. The main particulars of this ship [1] are as follow:— Length 130 ft., breadth 14 ft., displacement, including 19.48 tons load, 94.625 tons. The turbines developed about 1900 horse-power, and the vessel attained a mean speed during a two hours' trial of 26.205 knots, and a mean speed on the measured mile of 26.663 knots. The consumption of coal at full power was 2 tons per hour, and at 14 knots only 6 cwt. 3 qr. 8 lb. per hour. The turbines are arranged in the same way as the parallel-flow turbine installation in the "Turbinia," as shown in Plate XXXIV., adjoining page 74, except that a reversing turbine is embodied in the main low-pressure end, and a cruising turbine is also fitted on the centre shaft.

[1] See "ENGINEERING," vol. lxxviii., page 183.

France having led the way with foreign marine turbines, it will be convenient to describe the developments up to the present time in that country. The completion of the torpedo boat No. 293 was followed by a long pause, during which other countries regained the lead which France had originally obtained; but at the same time the little boat in question had been giving a good account of herself in service.

The next order was for one of the torpedo boat destroyers of the 1906 programme, named "Chasseur." The engines of this boat were the first warship turbines to be made in France, and a plan of the general arrangement of the machinery, as well as a view of the ship, are given on Plate LXXXII., adjoining this page. The "Chasseur" was completed in 1909, and has a length of 210 ft. 6 in., and a breadth of 22 ft. She had a displacement tonnage of 458 tons on trial, when, with a horse-power estimated at about 9000, she steamed 30.4 knots.

The machinery arrangement of the "Chasseur," as shown on Fig. 205, differs considerably from that of British destroyers of the same period: only one cruising turbine was fitted, and the machinery was arranged in compartments one abaft the other. The forward engine room contains the high-pressure turbine on the centre-shaft, and one low-pressure turbine on the starboard shaft, while its condenser is placed on the port side of the high-pressure turbine. In the after engine room there is the cruising turbine on the starboard shaft, and the other low-pressure turbine on the port shaft, the condenser being in the centre. The engines in both rooms are worked together normally, the turbines being handled from a single starting platform; but in the case of a breakdown in either engine room the vessel can be navigated with two shafts out of the three, or one engine room can be entirely abandoned, if filled with water. Were the forward compartment flooded the cruising turbine would, of course, suffice to drive the starboard shaft, and the port shaft would be rotated by the low-pressure turbine. These turbines were built by the Cie. Electro-Mécanique, of Le Bourget.

Late in 1906, after a long consideration by the Engineers of

Fig. 202. Steaming 26.20 Knots on Three Hours' Trial.

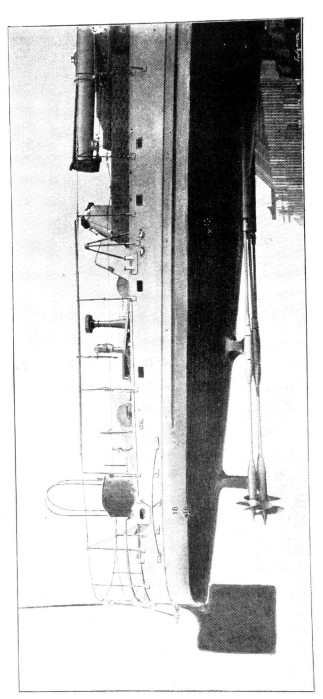

Fig. 203. The Propellers.

Plate LXXXI.—The First French Turbine-Driven Ship (1902) : Torpedo Boat No. 293.

Fig. 204.   Steaming 30.4 Knots.

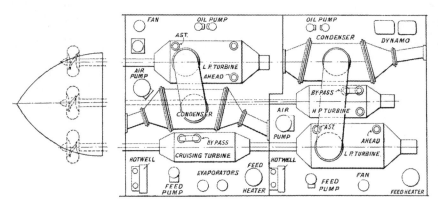

Fig. 205.   General Arrangement of Turbines.

**Plate LXXXII.—The French Torpedo=Boat Destroyer ''Chasseur''
and her Machinery.**

Fig. 206. Turbine Rotor.

Fig. 207. Rotors in Casings.

Plate LXXXIII.—The Turbines of the French Battleship "Voltaire."

Plate LXXXIV.—The French Battleship "Voltaire" (Fig. 208).

the Navy and a discussion in the Chamber of Deputies, it was decided that the six battleships of the Danton class, then to be laid down, should be fitted with Parsons turbines, although all the drawings had originally been prepared on the supposition that they would have reciprocating engines. This condition led to the necessity of fitting the turbines in exactly the same space and position in the ship as the proposed piston engines. These vessels, therefore, have the three turbine engine rooms amidships with boiler groups both forward and abaft of them. This arrangement permitted turbines of very large diameters to be fitted, as they were to be placed at the widest part of the vessel, and it was thus possible to make them very short; so that the space taken up in the fore and aft direction has been reduced to a minimum. On Plate LXXXIII., adjoining this page, there are two views of the turbines of the "Voltaire" in the La Seyne works of the Forges et Chantiers Company, where they were manufactured.

The centre engine room contains the two low-pressure turbines driving the inner shafts, as well as the high-pressure and inter-mediate-pressure cruising turbines and the manœuvring gear for all the turbines. The wing engine rooms each contain the high-pressure ahead and high-pressure astern turbines driving the wing shafts. Abaft these three rooms are two rooms for the condensers with the usual auxiliaries. So far as the turbines are concerned, therefore, it will be seen that the arrangement is similar to that of the "Dreadnought," with the exception of the two cruising turbines being in series.

The names of the six battleships are:—"Voltaire," "Danton," "Vergniaud," "Condorcet," "Diderot," and "Mirabeau." Four of the ships were built in private yards, and two in the Naval arsenals, the orders for the six sets of turbines being divided between the turbine shops at La Seyne and Saint-Nazaire. Each vessel has a length between perpendiculars of 475 ft., a breadth of 84 ft. 7 in., and on a mean draught of 27 ft. 1 in. the displacement tonnage is 18,374 tons. The designed speed is 19.25 knots, with about 22,500 shaft horse-power.

The "Voltaire" is the first of these battleships to be completed, and of her a view is given on Plate LXXXIV., facing page 127. Although the boiler power is about equivalent to that of the earlier ships of the Dreadnought class, it was not to be expected that so high a speed could be attained, because the ships are much shorter and broader. The "Voltaire," on a preliminary trial, developed 19,600 shaft horse-power, giving a speed of 19.3 knots. Two more battleships were laid down on August 1, 1910, and will be named the "Jean Bart" and "Courbet." They also are to be fitted with Parsons turbines. These vessels will be 542 ft. between perpendiculars by 88 ft. 9 in. beam, and at 28 ft. mean draft will have a displacement tonnage of 23,467 tons. The designed speed is 20 knots with 28,000 horse-power.

The laying down of the torpedo-boat destroyer "Chasseur" was also followed in 1907 by orders for three further destroyers—the "Janissaire," "Fantassin" and "Cavalier"—of which the turbines were identical with those of the "Chasseur," and the hulls very similar. A departure was, however, made by firing the boilers with liquid fuel instead of coal. These vessels developed speeds several knots in excess of the reciprocating-engined boats ordered at the same time.

The year 1908 is notable in the history of the French Navy as it was the first year in which no vessels were ordered with reciprocating engines. Two destroyers of that year named the "Bouclier" and "Casque" are enlarged editions of the Janissaire class, having displacements of between 650 and 700 tons. The boilers use liquid fuel, and the arrangement of Parsons turbines is similar to that in the "Chasseur," as shown on Plate LXXXII., adjoining page 126. The power, however, has been practically doubled and the contract speed is 31 knots.

In 1909 a further development took place in destroyer practice, all the boats of this year being fitted with twin screws. They are otherwise in all respects similar to the 1908 boats; and the same may be said of the boats of the 1910 programme. The type of turbine, of the Parsons combination impulse and reaction design, will be dealt with in our next chapter.

Since the introduction of the marine steam turbine, France

Plate LXXXV.—The First French Merchant Turbine Steamer: the "Charles Roux" (Fig. 209).

Plate LXXXVI.—The First Parsons Turbine made in France (1908) : Low-Pressure Turbine of the "Charles Roux" on Testing Bed (Fig. 210).

has not laid down any cruisers, either large or small. The last of the reciprocating-engined cruisers were completed in 1910.

The first marine turbines to be built in France were not, however, for a warship, but for the Mediterranean steamer " Charles Roux." This is a 19-knot steamer of 9000 horse-power which plies in the Marseilles-Algiers service of the Compagnie Générale Transatlantique, and illustrations of the ship and turbines are given on Plates LXXXV. and LXXXVI., facing pages 128 and 129 respectively. Her turbines were built by the Cie. Electro-Mécanique.

The St. Nazaire Company has built the Transatlantic liner " France," of the following dimensions :—Length between perpendiculars 688 ft., breadth 75 ft. 6 in., mean draught 29 ft., and displacement 25,460 tons; a sea speed of 23 knots is expected with about 40,000 horse-power. This is the largest and most powerful vessel ever built in France, and the turbines present several notable departures.

In addition to this large vessel, the same Company has on the stocks an intermediate liner, to be called the " Rochambeau "; and, as this is a vessel of low speed, the combination of reciprocating engines and turbines has been adopted. This arrangement of machinery is described in a later chapter on Developments in Merchant Ship Turbines. The " Rochambeau " is notable, as she is the first vessel with this type of machinery to have four shafts—the two inner shafts being driven by reciprocating engines and the wing shafts by exhaust steam turbines. This vessel has a length of 535 ft., a breadth of 62 ft. 9 in., and her displacement is 17,400 tons. The exhaust steam turbines will develop together about 4000 horse-power, and, with the combined power of reciprocating and turbine engines, the ship will have a speed of about 16 knots.

There is also building in France a Channel steamer for the Newhaven-Dieppe service, to be known as the " Newhaven." She is a similar vessel to the " Dieppe " described on page 115, and the turbine system of propelling machinery was adopted as a direct result of the efficiency of the " Dieppe." The " Newhaven " will be of slightly increased dimensions, and the machinery will be of rather greater power. The " Newhaven," too, differs, in that she is being fitted with Belleville steam generators instead of cylindrical boilers.

S

GERMANY.

In Germany the marine turbine was originally regarded with the greatest scepticism, and the builders of warships (especially of destroyers) waged an energetic campaign on behalf of the reciprocating engine. It was not until late in 1902 that the order for the first turbine destroyer was finally decided on. This boat, the S. 125, was one of the series of six boats ordered in that year, the remaining boats of the series having the ordinary type of reciprocating engines. This vessel, S. 125, which is illustrated on Plate LXXXVII., facing this page, has a length of 210 ft., a breadth of 22 ft., and a displacement tonnage of about 413 tons.

The trial results of this boat gave rise to a great amount of discussion, in which strenuous efforts were made to deduce the superiority of the reciprocating engine; but all such discussion was silenced by the results in service, which were very conclusively in favour of the turbine boat. As regards the type of machinery fitted, the engines of the S. 125 are practically identical in design and arrangement with those of the British destroyer "Eden," which was ordered in the same year, and is described on page 82. Two views of the turbines of the torpedo boat S. 125 are given on Plate LXXXVII.

In 1903 the small cruiser "Lübeck" was ordered to be built, and Parsons turbines were included in the design, so that comparison could be made with the performances of a series of similar vessels having reciprocating engines, the "Hamburg" being chosen as the standard ship. This comparison between the "Hamburg" and the "Lübeck" may further be extended by comparing both of them with the "Amethyst," described on page 84. All of these ships were constructed at the same time, and are of similar dimensions, displacement, and engine power.

The machinery arrangement in the "Lübeck" differs from that in the "Amethyst," as the former has four shafts instead of three. She was the first cruiser thus engined. This arrangement was necessitated by the introduction of a middle-line bulkhead as specified. The shafts on each side of the ship are driven by a high-pressure

Fig. 211.   The Torpedo Boat S. 125.

Fig. 212.   Cruising and High-Pressure Turbines.

Fig. 213.   The Turbines in Series.

Plate LXXXVII.—The First German Vessel with Parsons Turbines (1902):
The Torpedo Boat S 125.

Plate LXXXVIII.—The German Cruiser "Lübeck" (Fig. 214).

and a low-pressure turbine, and there are also two cruising turbines arranged in series with each other and with the high-pressure turbines. An independent high-pressure reversing turbine was fitted on each outer shaft, and a similar high-pressure reversing turbine was embodied in each low-pressure turbine. This latter arrangement has not since been repeated, and such reversing turbines are now always worked in series.

A comparative Table of dimensions of the vessels is given below.

TABLE VIII.—DIMENSIONS OF GERMAN AND BRITISH CRUISERS.

———	"Lübeck" and "Hamburg."	"Amethyst."
Length between perpendiculars...	340 ft.	360 ft.
Breadth	42 ft. 4 in.	40 ft.
Mean draft	15 ft. 9 in.	14 ft. 6 in.
Displacement	3150 tons	3000 tons
Boilers—Type	Schultz	Yarrow
Heating surface	29,800 sq. ft.	25,968 sq. ft.
Grate area	585 sq. ft.	493.5 sq. ft.

It will be seen that the British cruiser is slightly longer, narrower, and lighter, but with a little less boiler power than the German ships. A view of the "Lübeck" is given on Plate LXXXVIII., facing this page.

In the German ships much greater stress was laid on the radius of action at low cruising speeds. Something therefore was sacrificed at the top speed. The result is that the "Amethyst" turned out a faster ship than either the "Lübeck" or the "Hamburg," but both these vessels have considerably larger radius of action at low speed than the British ship.

Exhaustive trials were made with tandem propellers on each shaft of the "Lübeck," as compared with single ones. The turbine machinery, having been designed to work with tandem propellers, gave the best all-round efficiency at most speeds; but single propellers were finally preferred, owing to considerations of simplicity, better manœuvring qualities, and a slightly better performance at top speed.

The comparative miles run per ton of coal at various speeds by the four ships whose dimensions are given in the Table on page 131 are plotted on the curve Fig. 215, on this page. This comparison is especially interesting. The performance of the "Hamburg" must be looked upon as remarkable for reciprocating engines, but that of the "Amethyst" shows the highest efficiency of any of the vessels at the high speeds. In this comparison it should be noted that the measure of efficiency includes hulls and boilers, as well as engines and screw propellers, and is not one purely of engines.

The results obtained by the "Lübeck" and the "Hamburg" in service showed figures more favourable to the turbine ship than on the comparative official trials, the "Lübeck" being much the more economical. It was not long, therefore, before a second small turbine-driven cruiser, the

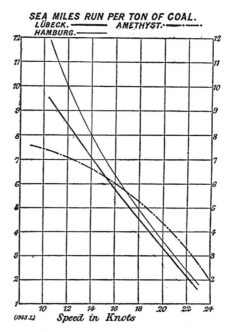

Fig. 215. Comparative Results of Coal Consumption

of German Cruiser "Hamburg" with Piston Engines, and "Lübeck" with Turbine Engines, and of British Cruiser "Amethyst" with Turbine Engines.

"Stettin," was ordered.   This vessel attained a speed of 25 knots on her continuous full-power trial, and was shortly followed by the "Dresden," which did equally well, and is notable as the first turbine warship to use superheated steam.   The "Dresden" is illustrated on Plate LXXXIX., facing this page.

All these small cruisers had the same arrangement of turbines as the "Lübeck," with the exception that the reversing turbines were in series.   A fourth vessel of this type—the "Augsburg"—

Fig. 216.   The German Cruiser "Dresden."

Fig. 217.   The German Torpedo-Boat Destroyer G 137.

Plate LXXXIX.—Recent German Ships with Turbine Machinery.

Plate XC.—The German Armoured Cruiser "Von der Tann" (Fig. 218).

is the first vessel for which turbines were built in one of the Imperial Dockyards. A fifth small cruiser, the "Ersatz Condor," has been ordered, and her turbines will also be built in an Imperial Dockyard.

Meanwhile, an order for a new torpedo-boat destroyer—G.137— had followed shortly on the completion of the S.125. This vessel, of which an illustration is given on Plate LXXXIX., has a length of 233 ft., a breadth of 24 ft. 7 in., and a displacement tonnage of 570 tons. She was the most powerful destroyer built up to that time in Germany, and had nearly double the power of S.125. The machinery is arranged in two engine-rooms, one abaft the other, with coal protection at the sides. There is one cruising, one high-pressure, and two low-pressure turbines, the high-pressure and one low-pressure being in the forward room, and the cruising and the other low-pressure in the after room. Although designed for a speed of 30 knots she attained 33.08 knots on a three hours' run, and may be said to have sealed the fate of the reciprocating engine for destroyer work in Germany, because all destroyers ordered subsequent to her trials have been fitted with turbines. Four more turbine boats of the same power, with three screws, have followed the G.137, and twin screws have been adopted for all later boats. All the destroyer turbines, as well as those of the "Lübeck," "Stettin," and "Ersatz Hagen," were supplied by the German Turbinia Company, of Berlin, and were built in the Mannheim Works of Messrs. Brown, Boveri, and Co.

The first large armoured vessel built in Germany with turbines is the cruiser "Von der Tann,"[1] which has the following dimensions :—Length 561 ft., breadth 85 ft. 7 in., and on a mean draught of 26 ft. 7 in. the displacement tonnage is about 19,000 tons. The "Von der Tann" is illustrated on Plate XC., facing this page.

This vessel is fitted with cruising turbines, and the machinery partakes of the general characteristics of large warship turbines of the date when she was laid down. Three further vessels of the same type are under construction—the "Moltke," which was

[1] See "ENGINEERING," vol. xc., page 27.

launched in the spring of 1910, and the "H" and "J" which are on the stocks at the date of writing. All are being fitted with Parsons turbines. The trials of the "Von der Tann" proved thoroughly satisfactory, the speed attained being over 27 knots, with about 70,000 shaft horse-power. All these cruisers, as well as their turbines, have been constructed by Messrs. Blohm and Voss, Hamburg.

The last type of vessels in which the German Admiralty introduced turbines were the battleships, the delay in their adoption in such ships being no doubt partly due to the necessity of preserving homogeneity in the classes of battleships already commenced with reciprocating engines. Two ships were laid down in 1909 to replace the "Hildebrand" and "Hagen," and the type of turbine fitted in these vessels shows some departures which are at present special to the German Navy, both vessels being fitted with triple screws.

So valuable has the distinctive Parsons system of arranging turbines in series been proved, that this arrangement is being adopted in several turbine vessels now building in Germany. This is the case both with regard to battleships and small cruisers, and the latest instance is the very large Hamburg-Amerika liner, being built at the Hamburg Works of the Vulcan Company of Stettin. In this ship the turbines are to be arranged in series under special license, although they are not of the acknowledged Parsons type.

This vessel will, with the exception of the experimental Curtis-A.E.G.-turbined passenger boat "Kaiser," built in 1905, be the first turbine-driven merchant vessel in Germany—a very remarkable circumstance when the importance of her mercantile marine is remembered. It is the case, however, that the number of fast merchant vessels built in Germany since the introduction of steam turbines has been extremely limited. No doubt the immediate future will see a considerable change in this respect.

UNITED STATES OF AMERICA.

The first vessel to be fitted with Parsons turbines in the United States was the "Governor Cobb," a vessel built by Messrs. W.

Plate XCI.—The United States Cruiser "Chester" (Fig. 210).

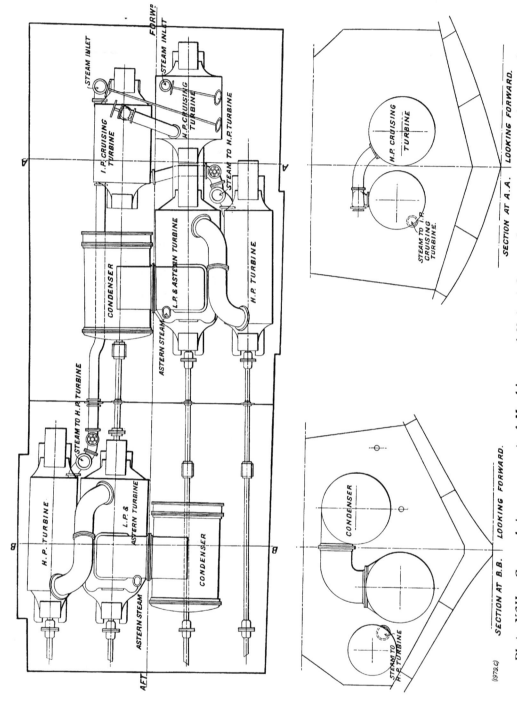

(979.C)

Plate XCII.—General Arrangement of Machinery of United States Cruiser "Chester" (Fig. 220).

SECTION AT A.A. LOOKING FORWARD.

SECTION AT B.B. LOOKING FORWARD.

and A. Fletcher Company for the Eastern Steam Ship Company in 1905 for service between Boston and St. John's. This vessel has a length of 290 ft., a beam of 51 ft., and, on a loaded draft of 14 ft., the displacement is about 2900 tons. The arrangement of machinery is identical to that adopted in the " King Edward " —described on page 99—and subsequent Channel steamers, there being one high-pressure and two low-pressure turbines, each driving a separate shaft, with an astern turbine in the exhaust casing of each of the low-pressure turbines. The contract speed on service was 17½ knots, for which the estimated equivalent indicated horse-power was 4000. On her trial in October, 1906, the speed attained was 20.5 knots.

At about the same time as the " Governor Cobb " was ordered, the United States Government placed the contract for the three scout cruisers, the " Birmingham," with reciprocating engines, the " Salem," with Curtis turbines, and the " Chester," with Parsons turbines. The Bath Iron Works were entrusted with the construction of the scout cruiser " Chester." The turbines in the " Chester " were arranged on four shafts in two engine rooms, with a transverse bulkhead between the two. Cruising turbines in series were fitted, one on each of the low-pressure turbine shafts. Astern turbines were only fitted on two of the four shafts, being incorporated in the exhaust casings of the low-pressure turbines. A view of the " Chester " is given on Plate XCI., and the arrangement of machinery is shown on Plate XCII., facing this page.

These three vessels formed the basis for a very interesting comparison, inasmuch as they were identical, with the exception of the type of propelling machinery. They were each 420 ft. long, and on a draught of 16 ft. 9 in. displaced about 3750 tons. The contract speed was 24 knots, and on the full power trial the " Chester " obtained a speed of 26.52 knots, the " Salem " 25.947 knots with practically the same total coal consumption, and the " Birmingham " 24.325 knots. The comparative results on trial and in service have been repeatedly reviewed.[1] The only commentary

---

[1] See "ENGINEERING," vol. lxxxvi., page 312, regarding comparative results; also "ENGINEERING," vol. lxxxix, page 537.

note which need here be made is that on the conclusion of the trials the United States Government ordered Parsons machinery for their new battleships, as will presently be stated.

The next vessels to be designed with Parsons turbines were the " Yale" and " Harvard" for which the Metropolitan Steam Ship Company placed the order with Messrs. W. and A. Fletcher Company about the end of July, 1905. They are engaged on passenger service between Boston and New York, and have a length of 386 ft., a beam of 50 ft., and, on a loaded draft of 16 ft., the displacement is about 4600 tons. The designed speed was 20 knots, with an equivalent indicated horse-power of 10,000, and on the official trials in 1907 they attained a speed of $22\frac{1}{2}$ knots.

These vessels were closely followed by the " Old Colony," the " Camden" and the " Belfast." The " Old Colony" was built for the New England Steam Ship Company, for service between Boston and New York. The keel was laid at the end of 1905 at the works of Messrs. W. Cramp and Sons, Philadelphia, and the turbine machinery was made by the Quintard Iron Works and Messrs. W. Cramp and Sons. The vessel is 375 ft. long and displaces 3800 tons when at a draught of 14 ft. 6 in. On trials at the end of 1907 the vessel attained a speed of 20.04 knots, or $1\frac{1}{2}$ knots in excess of the contract speed.

The " Camden" and " Belfast" were built and engined by the Bath Iron Works, Limited, for the Eastern Steam Ship Company for service between Boston and Bangor. These were smaller vessels, being 320 ft. long and of 1820 tons displacement on a draught of 9 ft. 3 in. The " Camden" attained a speed of 19 knots, being two knots in excess of the guaranteed speed, and the " Belfast" a speed of $18\frac{1}{2}$ knots.

In September, 1907, the contracts for five destroyers with Parsons turbines were awarded by the United States Navy Department—the " Charles W. Flusser" and the " Samuel C. Reid" to the Bath Iron Works, Limited, the " Joseph B. Smith" and the " Roswell H. Lamson" to Messrs. W. Cramp and Sons, and the " Samuel W. Preston" to the New York Shipbuilding Company. These destroyers were 289 ft. in length, 26 ft. 5 in. in beam, and at

Fig. 221. United States Destroyer.

Fig. 222. General Arrangement of Machinery.

Plate XCIII.—United States Destroyer and her Machinery.

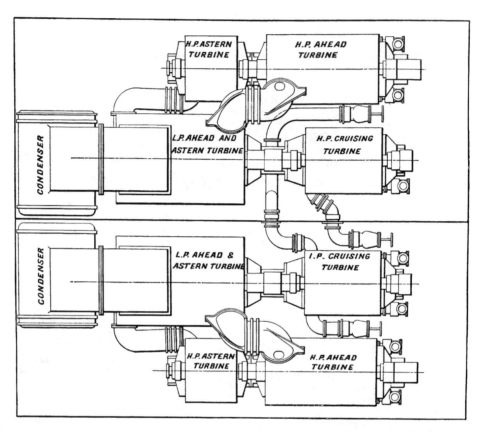

END VIEW LOOKING AFT.

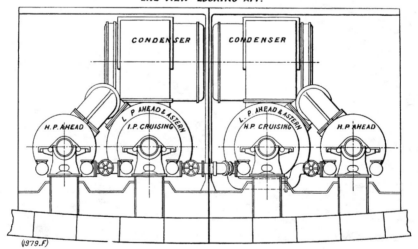

(1979.F)

Plate XCIV.—General Arrangement of Machinery in United States
Battleships (Fig. 223).

8 ft. ½ in. draught, the displacement was 700 tons. The contract speed was 28 knots. The machinery corresponds to that in the British torpedo-boat destroyers with three shafts and cruising turbines in series. The vessels are now in commission, having completed all their trials with very satisfactory results. The speeds, varying between 28.35 knots and 31.82 knots according to power developed, averaged 29.67 knots, while the steam consumption varied between 14.64 lb. and 18.84 lb. per shaft horse-power per hour for the main engines, averaging about 15 lb.[1]

Six destroyers of the 1908 programme were also fitted with the Parsons type of turbines, including the "Paulding" and "Drayton" by the Bath Iron Works, the "Roe," "Terry McCall" and "Burrow" by the Newport News Shipbuilding Company. The designed speed was 29½ knots, and this was exceeded in some cases by three nautical miles per hour. The destroyers "Ammon," "Monagan," "Patterson," and "Trippe," of the 1909 programme, bring the total number of destroyers for the United States Navy to date (1910), fitted with Parsons turbines, to 15. The arrangement of turbines in all of these destroyers is similar to that adopted by the British Admiralty, having the usual three-shaft arrangement with cruising turbines in series. A view of one of these United States torpedo-boat destroyers, and of the machinery erected in the contractor's workshop, is given on Plate XCIII., facing page 136.

After the exhaustive tests of the Parsons turbines in the scout cruiser "Chester," previously referred to, the United States Navy Department adopted the Parsons type for the four large battleships now building, the "Utah," "Florida," "Arkansas" and the "Wyoming." The "Utah" and "Arkansas" are being con-structed by the New York Shipbuilding Company, the "Florida" by the United States Navy Department, and the "Wyoming" by Messrs. W. Cramp and Sons. The "Utah" and "Florida" of the 1908 programme are of about 21,800 tons displacement and 28,000 horse-power, whilst the "Arkansas" and "Wyoming" are of 26,000 tons displacement and 30,000 horse-power.

[1] See "ENGINEERING," vol. lxxxix., page 538.

T

The arrangement of machinery in these vessels consists of a high and low-pressure turbine on each side of the centre line of the ship, each driving a separate shaft. There is also an astern turbine on each shaft, the high-pressure astern turbines being on the same shafts as the high-pressure ahead turbines, but each is in a separate casing, whilst the low-pressure astern and low-pressure ahead turbines are in the same casings. A complete set of engines on each side of the vessel is thus obtained. Cruising turbines in series are fitted, one on each of the low-pressure shafts. The general arrangement of the machinery is illustrated on Plate XCIV., facing page 137.

JAPAN.

With the rapid development of both the Imperial and Merchant fleets of Japan it was natural that great attention should be given to the subject of propulsion by turbine machinery. Four merchant vessels were fitted with turbines in 1908. Two of these, the "Tenyo Maru" and the "Chiyo Maru," were built at the Mitsu Bishi Dockyard, at Nagasaki, but were engined by The Parsons Marine Steam Turbine Company, Limited. Two others, the "Hirafu Maru" and the "Tamura Maru," were built and engined by Messrs. Denny, of Dumbarton.

The machinery of the "Tenyo Maru" and "Chiyo Maru,"[1] illustrate the progress achieved in the design of turbines for moderate sized mail liners, and engravings of the ship and machinery are given on Plates XCV. to XCVIII., between this page and page 139. The design of the turbines adopted in H.M.S. "Dreadnought" was generally followed in the case of these vessels, with the exception that the bearing ends were cast quite separately from the main body of the casings and were bolted on, the main body top and bottom halves being cast in one piece for the high-pressure, and in two pieces for the low-pressure, turbines. This reduced the number of vertical joints about the casings to a minimum, consistent with the important aim of keeping the castings within practicable dimensions. The turbines of these

See "ENGINEERING," vol. lxxxvi., page 592.

Plate XCV.—The Japanese Pacific Liner "Tenyo Maru" (Fig. 224).

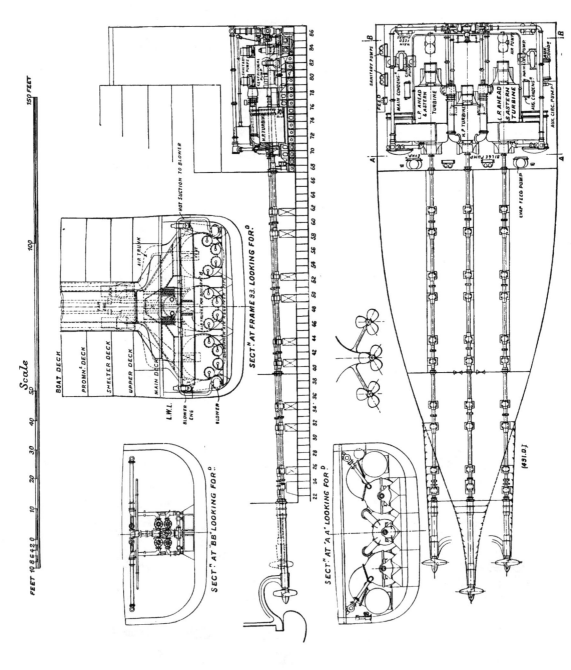

Plate XCVI.—General Arrangement of Machinery of "Tenyo Maru" (Fig. 225).

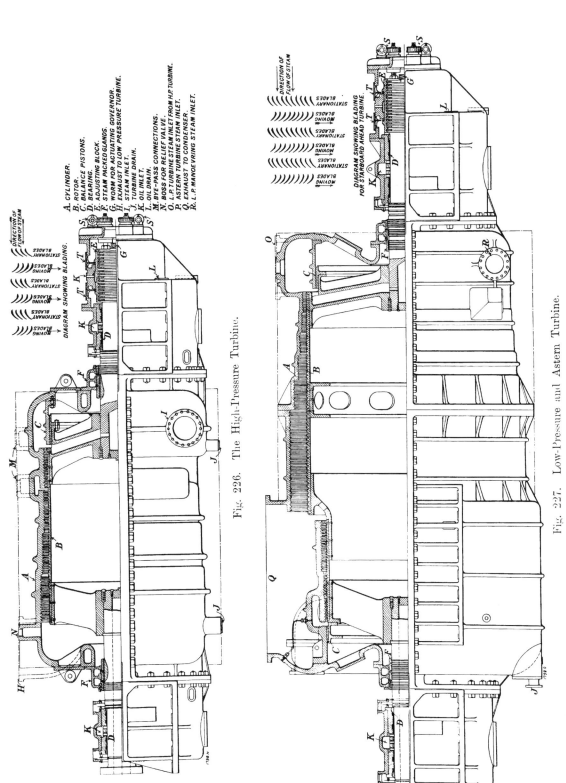

A. CYLINDER.
B. ROTOR.
C. BALANCE PISTONS.
D. BEARING.
E. ADJUSTING BLOCK.
F. STEAM PACKED GLANDS.
G. WORM FOR ACTUATING GOVERNOR.
H. EXHAUST TO LOW PRESSURE TURBINE.
I. STEAM INLET.
J. TURBINE DRAIN.
K. OIL INLET.
L. OIL DRAIN.
M. BYE-PASS CONNECTIONS.
N. BOSS FOR RELIEF VALVE.
O. L.P. TURBINE STEAM INLET FROM H.P. TURBINE.
P. ASTERN TURBINE STEAM INLET.
Q. EXHAUST TO CONDENSER.
R. L.P. MANOEUVRING STEAM INLET.

DIRECTION OF FLOW OF STEAM

STATIONARY BLADES
MOVING BLADES
STATIONARY BLADES
MOVING BLADES
STATIONARY BLADES
MOVING BLADES

DIAGRAM SHOWING BLADING.

Fig. 226.   The High-Pressure Turbine.

DIRECTION OF FLOW OF STEAM

STATIONARY BLADES
MOVING BLADES
STATIONARY BLADES
MOVING BLADES
STATIONARY BLADES
MOVING BLADES

DIAGRAM SHOWING BLADING
FOR STARBOARD AHEAD TURBINE.

Fig. 227.   Low-Pressure and Astern Turbine.

Plate XCVII.—The Turbines of the "Tenyo Maru."

Fig. 228. The High-Pressure and One of the Low-Pressure and Astern Turbines.

Plate XCVIII.—The Turbines of the "Tenyo Maru."

vessels are illustrated in section on Plate XCVII., and in perspective, from the forward end, on Plate XCVIII.

Some little trouble having been experienced with the astern dummies of the turbines in one or two vessels previous to this date, the fin packing having accidentally fouled and rolled over and thereby increased the clearance, the astern dummies in this and subsequent vessels were made with a strong vertical flange to minimise the possibility of the drum expanding. In order to meet the increase in the size of the thrust adjusting blocks, and to obviate the lifting of large blocks (or keeps) and the dismantling of the oil system when an adjustment was required, the top and bottom blocks were arranged to slide within the casting and the top keep, and means were provided by worm and gear (S on Figs. 226 and 227) to move these blocks independently from the exterior. Small doors on the top of the thrust block keeps (T on Figs. 226 and 227) permitted the usual liners to be inserted after the position of the rotor had been regulated.

The diameter of the high-pressure rotor is 80 in., and the blade heights ranges from $1\frac{5}{8}$ in. to $4\frac{1}{2}$ in. The overall length of this turbine is 23 ft., and the extreme diameter 9 ft. The diameter of the low-pressure rotor is 110 in., with blade heights ranging from $1\frac{5}{8}$ in. to 9 in. The astern rotor is 89 in. in diameter, with blades ranging from $\frac{5}{8}$ in. to $2\frac{1}{2}$ in. in height. The over-all length of the low-pressure ahead and the astern combined turbine is 31 ft. 6 in., and extreme diameter 11 ft. 6 in.

The volunteer steamers " Sakura Maru " and " Umegaka Maru "[1] were built and engined in 1908-1909 by the Mitsu Bishi Company, of Nagasaki, for the Imperial Marine Association. These vessels are 335 ft. in length by 43 ft. beam, and on a draught of 14 ft. have a displacement of about 2900 tons. The " Sakura Maru " attained a speed of 21.39 knots, and the "Umegaka Maru" of 21.3 knots. The arrangement of turbines in these vessels included the usual three shafts, one with the high-pressure and two with the low-pressure turbines.

The " Mogami," a despatch vessel for the Imperial Japanese Navy, was completed in the summer of 1908. This vessel, which was

[1] See " ENGINEERING," vol. lxxxviii., page 633.

also built by the Mitsu Bishi Company, was fitted with turbines built by The Parsons Marine Steam Turbine Company, Limited, Wallsend. This vessel, 300 ft. long, has also the three-shaft arrangement of turbines, and on a four hours' trial attained a speed of 23.18 knots.

The Mitsu Bishi Company have at present under construction Parsons turbines for a third vessel for the Toyo Kisen Kaisha, of the same dimensions as the "Tenyo Maru" and the "Chiyo Maru," and for a cruiser and two high-speed destroyers for the Imperial Japanese Navy.

But the most important Japanese turbine-driven ship is that ordered in 1910 by the Japanese Government from the Vickers Company. This vessel will be between 27,000 and 28,000 tons displacement, of great fighting power and exceptionally high speed, the contract price being about two and a-half millions sterling. The turbines in this case will be of Parsons impulse-reaction type, described in the succeeding chapter, and driving four shafts. They will be manufactured by the Vickers Company.

SPAIN.

Spain merits prominent notice in view of the naval programme being carried out in the Royal Dockyards by the Sociedad Española de Construccion Naval, which is managed by an advisory committee representative of three prominent naval construction companies in Britain. The three battleships, "España," "Alfonso XIII.," and "Jaime I.," though vessels of moderate size (about 16,000 tons), are designed to attain a speed of $19\frac{1}{2}$ knots, and are equally remarkable for their great offensive and defensive qualities.[1] A view of the vessels as they will appear when finished is given on Plate XCIX.

The turbines, as shown in the plan on Plate C., facing page 141, are of a special design, arranged in two engine-rooms as in early ships of the British Dreadnought class. They have main high-pressure, intermediate-pressure, and two low-pressure turbines without any cruising turbines, thus giving more efficiency and a better result when running at full speed. The arrange-

See "ENGINEERING," vol. xc., page 77.

Plate XCIX.—The New Spanish Battleships (Fig. 229).

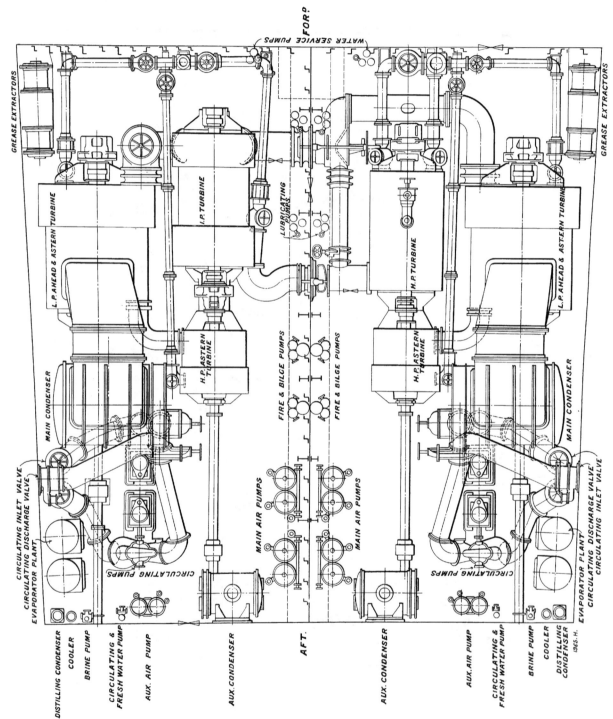

Plate C.—General Arrangement of Machinery in the Spanish Battleships (Fig. 230).

ment originally specified by the Spanish Government included cruising turbines, but was later modified in order to bring it as closely in accordance with the latest British practice as the guarantee named in the contract allowed.

The three destroyers — "Bustamente," "Villamil," and "Requesens" — are 380-ton vessels of 28 knots speed, with machinery of similar type to that of the British turbine destroyers of all classes from the Eden to the Tribal class, *i.e.*, having three shafts with one high-pressure, two low-pressure, and two cruising turbines in series. All these vessels were ordered in 1909. In addition there are in hand ten torpedo-boats of 180 tons and 26 knots, with 3500 horse-power machinery of similar type to that in the British "coastal" (or Nos. 1—36) torpedo-boats, *i.e.*, a high-pressure and an intermediate-pressure turbine driving the wing shafts, with a low-pressure turbine on the centre shaft, as well as a cruising turbine, half the power being developed on the centre shaft and a quarter on each of the wing shafts. This arrangement is shown in Figs. 150 to 152 on Plate XLVIII., facing page 93.

RUSSIA.

Russia has on the stocks the four battleships "Gangut," "Poltava," "Sevastopol," and "Petropavlovsk." [1] For these vessels Parsons turbines of 42,000 shaft horse-power are being built. The turbines themselves are of similar type, and arranged similarly as regards bulkheads, to those described for the French battleships.

ITALY.

The Royal Italian armoured cruiser "San Marco," constructed by Messrs. Gio. Ansaldo, Armstrong, and Co., of Genoa, is the first turbine ship of this important maritime power. On her trials in July, 1910, she attained a speed of 23.7 knots, being nearly 2 miles per hour in excess of the speed attained by a sister vessel, the "San Georgio," fitted with reciprocating engines. This vessel, illustrated on Plate CI., facing page 142, was commenced in May, 1907, and is 429 ft. 9 in. in length, 68 ft. 9 in. beam, and, on a

[1] See "ENGINEERING," vol. lxxxix., page 652.

draught of 23 ft. 6 in. displaces about 10,000 tons. The "San Marco" has the four-shaft arrangement now adopted in most ships, with cruising turbines in series. On each side of the ship there is a cruising turbine and a high-pressure and a low-pressure ahead and astern turbine.

The "Citta di Catania" and "Citta di Palermo," whose turbines were built by Messrs. Gio. Ansaldo, Armstrong, and Co., for the Italian State Railway Company's service between Naples and Palermo, have recently been put on service. These vessels are 340 ft. in length, 42 ft. 3 in. beam, and about 3500 tons displacement. On trial they attained speeds of 23 knots.

At the present time, there are under construction in Italy, for the Royal Italian Navy, Parsons turbines for four large battleships, a scout cruiser, several torpedo-boat destroyers, and a torpedo boat.

AUSTRIA.

It was not until 1907 that Austria ordered her first turbine cruiser, the "Admiral Spaun," a vessel of the scout class of the following dimensions:—length 427 ft., breadth 42 ft., displacement 3500 tons: with 25,100 shaft horse-power she has attained a speed of 27.07 knots. A view of the vessel completed is given on Plate CII., facing page 143. The turbines were built by the Stabilimento Tecnico of Trieste.

The battleships now building in Austria are being fitted with turbine machinery of the most modern type, and it is in contemplation to fit the forthcoming destroyers and torpedo boats also with turbines as soon as the present series with reciprocating engines are completed. On Plate CII. there is also a view of the turbines under construction for one of the battleships already referred to. These are likewise being built by the Stabilimento Tecnico.

SWEDEN AND DENMARK.

Sweden and Denmark have just ordered their first turbines for warships, a 7000-horse-power set for a 21-knot cruiser in the case of Sweden, and engines of 4000 horse-power for some 200-ton twin-screw torpedo boats in the case of Denmark.

Fig. 231. Ship on Trial.

Fig. 232. Turbines under Construction.

Plate CI.—The Italian Cruiser "San Marco."

Fig. 233. The Cruiser "Admiral Spaun."

Fig. 234. Turbines of the Battleship "Tegetthof."

Plate CII.—Austrian Turbine=Driven Warships.

BELGIUM.

Belgium does not possess a war navy, but three turbine vessels have been built for the Government passenger line between Ostend and Dover. The first of these, the "Princesse Elisabeth," possessed at the date of her completion in 1905 the record of speed for passenger vessels, 24.03 knots. This vessel was fitted with turbines by The Parsons Marine Steam Turbine Company, Limited, and in service she made the voyage in 15 per cent. less time than the paddle steamers built a year or two previously, although the consumption of coal was the same. For the same displacement and speed the reduction in coal consumption works out at 25 per cent. The two new vessels just completed, "Jan Breydel" and "Pieter de Coninck,"[1] have exceeded this speed by three-tenths of a knot, though the record has in the meanwhile been broken by the "Ben-my-Chree" with $25\frac{1}{2}$ knots (see page 118), and the "Lusitania" and "Mauretania" with 26 knots (see page 123). The turbines of the two later Belgian boats, as well as the hulls of all three, were constructed by the John Cockerill Company, of Seraing. The "Princesse Elisabeth" is illustrated on Plate CIII., facing page 144, and the "Jan Breydel" on Plate CIV., facing page 145.

BRAZIL.

There has been in recent years a very considerable development of the navies of the South American Republics, and it was natural that the turbine system should appeal directly to the authorities responsible for the acceptance of designs. Brazil embarked on a very considerable programme some years ago, and two of the most notable vessels—the scout cruisers "Bahia" and "Rio Grande do Sul"—are fitted with turbines, and both ships have proved themselves exceptionally efficient and economical, the average speed being $27\frac{1}{4}$ knots. This was exceedingly favourable, as the vessels are not much larger than the British scout cruisers built a few years ago, and of speeds between 25 and 26 knots. The Brazilian scouts, which were built by Sir W. G. Armstrong, Whitworth, and Co., Limited, Elswick, have a length over all

See "ENGINEERING," vol. xc., page 84.

of 401 ft. 6 in., or between perpendiculars of 380 ft. The breadth moulded is 39 ft., and at a mean draught of 13 ft. 6 in. the displacement tonnage is 3100 tons. A view of one of these vessels is given on Plate CV., facing page 146.

The machinery was constructed by the Vickers Company at Barrow-in-Furness, and, as shown in the drawings reproduced on Plate CVI., facing page 147, the turbines are arranged on three lines of shaft in one engine room. The high-pressure ahead turbine is on the centre shaft, and there is a low-pressure turbine on each wing shaft. At the forward end of each of the latter there is a cruising turbine, the port and starboard cruising turbines working in series, while at the after end there are astern turbines, so that the ship is manœuvred entirely by the wing propellers; the high-pressure turbine on such occasion runs *in vacuo*. The fitting of cruising turbines in series enables the high-pressure cruising turbine to be cut out at intermediate speeds, the steam from the boiler being then passed to the second or intermediate-pressure cruising turbine. The condensers are arranged in the wings of the ship, with the pumps, &c., contiguous to them. The boilers are of the Yarrow type—ten in number—and work at a pressure of 250 lb.

The "Bahia" attained a speed at full power of 27.016 knots, with the turbine running at 496.2 revolutions, while the pressure at the turbine receiver was 182 lb., and the vacuum 27.4 in. The second ship, the "Rio Grande do Sul," improved even upon these results, the steam pressure at the turbine being 190 lb., and the revolutions 507.1, which gave a speed of 27.412 knots. A notable feature is that the coal consumption, at about three-quarters power, when the speed was 24 knots, was in the "Bahia" only 1.54 lb., and in the "Rio Grande do Sul" 1.79 lb., per shaft horse-power per hour.[1]

The success of these two ships induced the Brazilian naval authorities to fit turbine machinery in their later ships, including the battleship "Rio de Janeiro," being built at Elswick, and being engined by the Vickers Company. This battleship, which promises to be the most powerful from the standpoint of fighting efficiency

[1] See *The Engineer*, vol. cix., page 514.

Plate CIII.—The Belgian Government's Ostend and Dover Steamer "Princesse Elisabeth" Steaming 24.03 Knots (Fig. 235).

Plate CIV.—The Belgian Government's Ostend and Dover Steamer "Jan Breydel" Steaming 24.3 Knots (Fig. 236).

yet laid down, has a length over all of 690 ft. and a beam of 92 ft., the displacement tonnage at 26 ft. 6 in. draught being about 32,500 tons. The anticipation of the design is that a speed of about 22½ knots will be realised. The turbine machinery will be arranged in four units, with single propellers on each shaft. In this case it is probable that the turbines will be of the combined impulse and reaction type, a development which is described in the succeeding chapter.

ARGENTINE.

The Argentine Government are adopting Parsons turbines in their new destroyers, now being built by Messrs. Cammell Laird and Co., of Birkenhead. These promise to be most remarkable vessels. They are named the " San Luis," " Sante Fé," " Santiago," and " Tucuman." Their displacement at full load will be 1175 tons when the draught is 10 ft., but normally they will draw 8½ ft. of water, when the displacement will be 980 tons. The length over all is 293 ft., and between perpendiculars 285 ft., and the beam 27¾ ft. The machinery in this case also will be of the combined impulse and reaction type, and the power to be developed is close upon 20,000 horse-power with the turbines running at 600 revolutions, when it is anticipated that the speed of the vessels will be 32 knots. This for vessels of such limited dimensions will be a remarkable achievement.

PERU.

In alluding to the work done for the South American Republics, brief reference may be made to two passenger steamers, the " Huallaga " and the " Ucayali." These vessels were built for the Peruvian Steamship Company by Messrs. Cammell Laird and Co., Limited, for trading on the West Coast of South America. They are 360 ft. long by 46 ft. beam, and displace about 6000 tons. The machinery resembles that fitted to British Channel steamers, and on trial the vessels attained a speed of 19 and 19.11 knots respectively.

U

CHINA.

China also is embarking upon a comprehensive scheme for the reconstruction of the Navy, and two cruisers, now being built respectively by the Vickers Company and by Sir W. G. Armstrong, Whitworth and Co., will form the nucleus of the new fleet, and will be fitted with Parsons turbines. These ships are to serve also the purpose of training the officers and men for the new warships, and it was therefore important that they should embody the latest ideas. The Vickers-built vessel, a protected cruiser, is 330 ft. long between perpendiculars and 39 ft. 6 in. breadth moulded, the displacement tonnage being 2460 tons at a mean draught of 13 ft. A speed of 20 knots is anticipated. The turbine machinery is arranged on three lines of shafting. No separate cruising turbines are adopted in this installation, a special series of cruising blades being fitted instead at the forward end of the high-pressure turbine. In view of the educational function of the ship the steam-generating plant includes a combination of cylindrical and water-tube boilers arranged in two separate compartments.

THE COLONIES.

We may complete this review of machinery built for navies other than the British with a brief reference to the work being done for the Colonial fleets. It will be remembered that the patriotism of Australia and New Zealand found expression some time ago in an offer by both Dominions to present an armoured ship to the Mother Country, and this offer finally resulted in a definite scheme for fleets for the various outposts of empire. Thus it was decided that there should be a Pacific Fleet consisting of three units, each with an armoured cruiser, three second-class cruisers, six destroyers, and three submarine boats for the East Indies, Australia, and China Seas respectively. Canada agreed to organise a fleet to suit her own requirements. The armoured ships contributed by Australia and New Zealand will be the flagships of the Australian and China Sea units, while a cruiser of the Invincible class, built for the Home Navy, will take a corresponding place in the East Indian unit. The Australian and the New Zealand ships are being built

Plate CV.—The Brazilian Scout "Bahia" (Fig. 237).

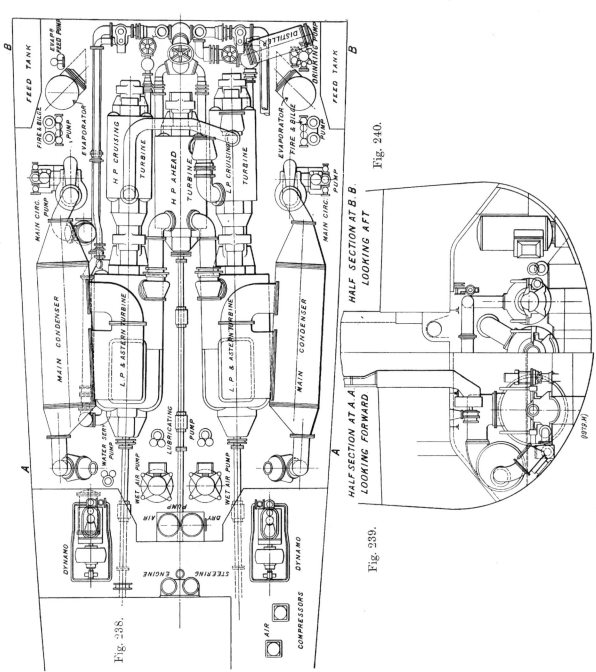

Plate CVI.—The General Arrangement of the Machinery of the Brazilian Scouts "Bahia" and "Rio Grande do Sul" (Figs. 238 to 240).

respectively by the Fairfield Shipbuilding and Engineering Company, Limited, and Messrs. John Brown and Co., Limited, and both are being fitted with Parsons turbine machinery, corresponding generally to that adopted in the Indefatigable and Invincible classes already referred to on page 91.

Two torpedo-boat destroyers have already been built for the Australian fleet, and have made a successful outward voyage, while the third has been sent across the seas in sections, to be finally completed in the Colony. All these vessels have Parsons turbine machinery corresponding generally with that fitted to the later British destroyers. They were designed by Professor J. Harvard Biles, of Glasgow University, and have a length between perpendiculars of 245 ft., a breadth, moulded, of 24 ft. 3 in., and a displacement of 700 tons. The Parsons turbines include a high-pressure and two low-pressure turbines for steaming ahead, with a corresponding arrangement for going astern. There are also high-pressure and intermediate cruising turbines. The turbines drive three shafts. The speed attained on an eight hours' trial by the "Paramatta," built by the Fairfield Company, Limited, was 26.869 knots, the turbines developing 11,212 shaft horse-power; while on the corresponding trial the "Yarra," built by Messrs. Denny, of Dumbarton, attained a mean speed of 27.08 knots. The mean of the two fastest runs was 28.5 knots, while the anticipation of the design was 26 knots.[1] This success admirably serves as a concluding reference to the work done with the Parsons reaction turbine in the British and foreign fleets, and paves the way for some investigation of the probable development in the immediate future—a consideration reserved for the succeeding chapter.

[1] See "ENGINEERING," vol. xc., page 404.

RECENT AND PROSPECTIVE DEVELOPMENTS IN WARSHIP TURBINES.

AN outstanding feature of the record in the three preceding chapters of the application of the steam turbine to ships in the British and Foreign Naval and Mercantile fleets is, it will be recognised, the fact that development has been along the lines laid down so definitely and completely in the first marine turbines constructed. This is only another evidence of the theoretical accuracy of the principles of the invention and of the mechanical efficiency of the application of those principles. Theory and practice were evolved as a consequence of protracted experimental research, and while there is abundant justification for contentment with the results, not only on account of the economy attained, but also from the full realisation of the most hopeful anticipations as to commercial success, in the adoption of the system for about five million shaft horse-power in ships and for an equally satisfactory total on land, yet the scientific skill and practical experience of those concerned with the invention are continually exercised in its further improvement.

It is a truism that there can be no finality in science, and a proof of the possession of the true scientific spirit is a continuous striving after higher success and wider utility. The present position of the Parsons turbine, and the attitude of those responsible for its extended application, completely exemplify these indisputable statements. It is thus important that the narrative of marine work in the preceding pages should be followed by some general review of recent and prospective developments. We will confine ourselves here to the wider issues and to warship turbines—leaving the mercantile marine and land work for consideration later—and refrain from reference to those improvements in detail already noted in connection with successive installations in warships or merchantmen.

In explanation of the causes which have led up to recent developments in warship turbines, we cannot do better than quote the remarks of Engineer Vice-Admiral Sir Henry Oram, K.C.B., in his Presidential Address to the Junior Institution of Engineers on November 16, 1909 [1] :—

" All our original turbine ships were fitted with cruising turbines in addition to those for use at the higher powers. The 'Dreadnought,' for example, has four turbines on each side of the ship, counting as one turbine the low-pressure and low-pressure astern, which are in one casing. Experience, however, has shown us that there are certain inconveniences attending the use of cruising turbines. They are economical, but being very often not in use they are apt to be neglected, and not to receive the attention they require, and the few accidents that have occurred with turbines have practically all occurred in the cruising turbines.

" The question arose as to whether the increased economy due to their use is worth the extra complication, cost, and liability to injury, and the conclusion is arrived at, at least in the case of single cruising turbines, that the balance of advantage and disadvantage is adverse to them, and they have not been fitted in recent warships. This view is the more readily taken as the alternative recommended by Mr. Parsons appears to give us most of what we require without multiplication of turbines. This alternative is considerably to increase the expansion allowed in the main turbines at high powers, provision being made to obtain the maximum power by means of by-pass arrangements. This ensures greater economy at low powers than was hitherto obtainable with the main turbines, and by this system we gain simplicity and also a reasonable economy at such low powers.

" In certain classes of ships, however, where the radius of action at low powers is exceedingly important owing to their limited coal or oil storage, as in torpedo-boat destroyers, two cruising turbines in series are fitted. In a typical example of this sort there are five turbines fitted on three shafts, again counting the low-pressure and

[1] See " ENGINEERING," vol. lxxxviii., page 703.

low-pressure astern turbine as one. In this case, the gain by the use of the cruising turbines in series is much greater, and is considered to make up for the extra cost, weight, and complication involved, and they are being retained."

The alternative referred to above by Sir Henry Oram consists in extending the main high-pressure turbine at its steam end by adding an additional stage or stages, so that when it is desired to run at lower fractions of power the full pressure drop of the steam can be utilised. But at higher powers, steam from the boiler may be added through by-pass valves fitted at intermediate stages. Thus, under all conditions of working, down to somewhat below half power, a satisfactory steam pressure is maintained. A turbine of this type is illustrated on Plate CVII., facing this page. A perspective view of another turbine of the same type — one of those fitted to H.M. cruiser " Newcastle "—is given on Plate CVIII., facing page 151.

At fractions of power much less than one half, however, the steam has still to be throttled at the point of admission, and this can never be regarded as equal to maintaining a high pressure for all fractions of power. The system does, however, give a reasonable economy at low powers, as stated by Sir Henry Oram, and it has been adopted to a considerable extent in large warships.

It must not, however, be lost sight of that the reduction in weight and space due to the suppression of the cruising turbines has been utilised to increase the efficiency of the main turbines; so that at full power they are considerably more efficient than those constructed during the period in which cruising turbines were still adopted.

The influence of this increased efficiency at full power is felt noticeably in the consumptions right down to about half power. Below half power, however, the efficiency of turbines of this type is undeniably somewhat less than that of the earlier combination with cruising turbines, such as is shown in the drawings reproduced on Plate XCIV., facing page 137, which illustrate a general arrangement of turbine machinery with cruising turbines suitable for four-shaft large war vessels.

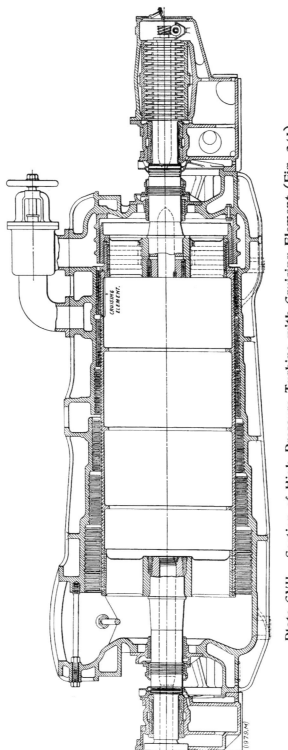

CRUISING
ELEMENT.

(1979.M)

Plate CVII.—Section of High=Pressure Turbine with Cruising Element (Fig. 241).

Plate CVIII.—Ahead High=Pressure Turbine with Cruising Element and By=Pass, with Separate High=Pressure Astern Turbine on the Same Shaft (Fig. 242).

More recently, however, the requirements of certain Admiralties with regard to radius of action at fractions so small as one-fifth or even one-tenth power call for the provision of some additional means to increase this radius of action and to attain a comparable economy without the use of separate cruising turbines.

Mr. Parsons, when working on the radial-flow turbine, as narrated in the preceding chapter on experiments in connection with the mechanical details of the turbine, was proceeding along the line leading to the partial-flow system by the use of nozzles and chambers—also, indeed, to the velocity-compounded impulse turbine; but when he recovered the rights in his reaction turbine patents he discontinued his experiments with the radial-flow turbine. The growing success of the parallel-flow turbine from 1894 onwards required concentration of effort in design work still further to increase efficiency at wide ranges of speed, and thus the time available for experimental research in other directions was limited.

With steam admission to the full area of the annular space between the rotor and casing occupied by the blading it was recognised that leakage losses at low speed were serious. To reduce these, the diameter and blade speed have been cut down at the high-pressure end, necessitating the use of many rows of blades to abstract the work from the steam, and even then the ratio of the blade speed to steam velocity has been lower than it should be for maximum economy. With partial admission the diameter at the high-pressure end may often be as great as at the low-pressure end, so that relatively few rows of moving blades are needed to abstract the energy of the steam, and the ratio of blade speed to steam speed may be very close to that required for maximum economy. The difficulty of adapting the system of partial or variable admission to the reaction turbine was great, owing to the fact that the steam falls in pressure as it passes through the moving blades. It is essential to the reaction principle that a pressure difference shall be maintained between the admission and the discharge side of the moving blades. It will not do therefore merely to direct a jet of steam on to an open wheel, as when driving an impulse wheel by partial admission. In this latter case there is

practically no difference in pressure between the opposite sides of
the moving blades, or, in other words, the steam pressure in the
clearance space between the nozzles and the moving blades is the
same as at the discharge side of the moving blades, and hence
there is no tendency to a lateral
spreading of the steam in the
clearance space.

With the reaction principle, on
the other hand, an excess pressure
in the clearance is essential, and
hence the jet must in some way be
prevented from spreading laterally
if partial admission is to be used.
Mr. Parsons' solution of this problem, which may be accepted as the
second alternative, in point of time
at least, is represented diagrammatically in our illustrations on
Plates CIX. and CX., facing pages
152 and 153. The turbine represented by Fig. 243 was intended for
driving electric generators, and the
low-pressure end was built on the
usual lines. The high-pressure end,
on the other hand, is entirely different, and consists of three cells,
in each of which is a drum carrying
a few rows of reaction blading.
Similar rows are fixed on the casing,
but, as shown in Figs. 245 and 246

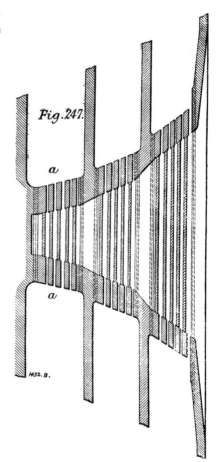

Fig. 247. Detail of Parsons Partial-
Admission Reaction Turbine.

on Plate CX., and in Fig. 247 on this page, these fixed rows do not
extend around the whole circumference of the shell, but are stopped
off as indicated. The steam enters the cell through the first row of
fixed blading, and its pressure at issue is substantially higher than
it is in the space $a\,a$ on Fig. 247. In addition, therefore, to flowing
through the moving blades opposite its point of entrance, it also

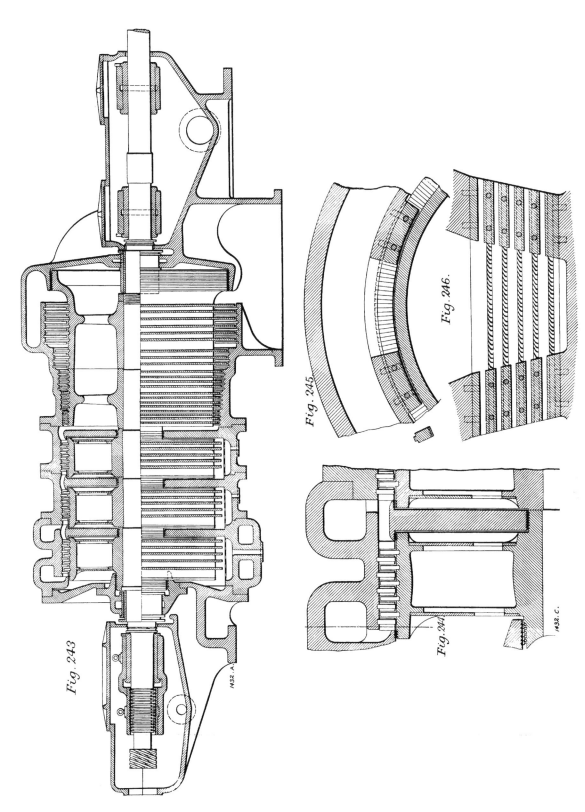

Fig. 243

Fig. 244.

Fig. 245.

Fig. 246.

Plate CIX.—Parsons Partial=Admission Reaction Turbine (Figs. 243 to 246).

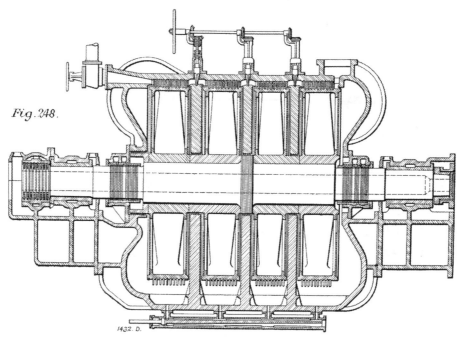

Fig. 248.

1432. D.

1978

Fig. 249.

Plate CX.—Parsons Partial-Admission Marine Turbine (Figs. 248 and 249).

leaks sideways into the space $a\ a$. Flowing on through the blades the steam continues to fall in pressure, and when about midway between the entrance and the exit from the cell it is at the same pressure as in the space $a\ a$. Hence during the second portion of its progress through the cell leakage takes place into the blades instead of out from them. An enlarged view of a cell showing the labyrinth packings by which leakage from cell to cell is minimised, is shown in Fig. 244. End balance is provided for in the fact that the labyrinth on the exhaust end of the boss is smaller in diameter than that at the forward end. As the working blades subtend only a small proportion of the whole circumference the total end thrust to be provided for is small. For balancing, the low-pressure end is provided with the dummy piston shown to the right in Fig. 243.

An application of the new construction for the cruising turbine of a warship is represented on Plate CX., facing this page. The perspective view on the same plate, with the top half of the casing removed, shows very clearly the arrangement of chambers already described. As will be seen, provision is here made for by-passing the steam direct to each cell in succession as the demand on the turbine increases. The section also shows the provision made for the draining of the cells.

A turbine of this type was constructed of dimensions to enable it to take the place of a high-pressure cruising turbine of an already existing set of machinery of a third-class cruiser, with the result of which, consequently, its output could be compared. Table IX. on the next page gives a comparison of the power obtained on test with the power developed by the existing high-pressure cruising turbine at the same revolutions.

From this Table of results it will be seen that the power has been increased by about 44 per cent., with approximately the same conditions of initial and final pressure. Since, however, the high-pressure cruising turbine when in use generates about 15 per cent. of the total power developed by all the turbines in the ship, it follows that the use of the new turbine in such an installation would lead to an increase of about 7 per cent. in the total power with the same quantity of steam at low speeds. The use of this type of partial-flow turbine is not limited to high-pressure cruising turbines. It can sometimes be advantageously employed at the high-pressure end of a main turbine, followed by a drum construction of the ordinary type.

TABLE IX.—COMPARISON OF RESULTS OF CRUISING TURBINE AND OF PARTIAL-ADMISSION TURBINE.

—	High-Pressure Cruising Turbine.	Partial-Admission Turbine.	
Initial pressure (pounds absolute) ...	230	140	195
Final pressure (pounds absolute) ...	80	55	78
Quantity of steam per hour (pounds) ...	33,000	24,200	33,000
Shaft horse-power...	240	248	345
Revolutions per minute	290	290	290

We come next to the third alternative, viz., the combination of one or more velocity-compounded impulse stages at the high-pressure end of a turbine with reaction blading for succeeding stages, in which case the volume of steam supplied to the turbine can be controlled by closing down as many of the admission nozzles in the impulse stage as may be required to suit the proportion of full power desired. By means of this combination the pressure drop which produces expansion of the steam in the first stage is confined to the fixed jets, and there is thus no loss by leakage. Drawings of a turbine of this type are reproduced on Plate CXI., facing this page.

The impulse type of blading with velocity compounding can in most cases be arranged to expand steam with fair efficiency to twice its volume in a single stage with blade speeds such as are permissible in marine work. It has not been considered advisable to use more than one such stage before going over to the simpler, cheaper, and more efficient reaction blading with drum construction. In the latest type of Parsons turbine the disc-and-drum construction can thus be used exclusively, so that all necessity for diaphragms

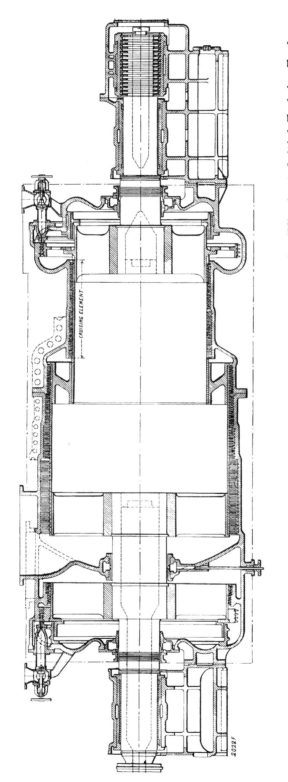

CRUISING ELEMENT

2022F

Plate CXI.—Section of High=Pressure Ahead and Astern Turbines with Impulse Wheel at Initial End in Each Case (Fig. 250).

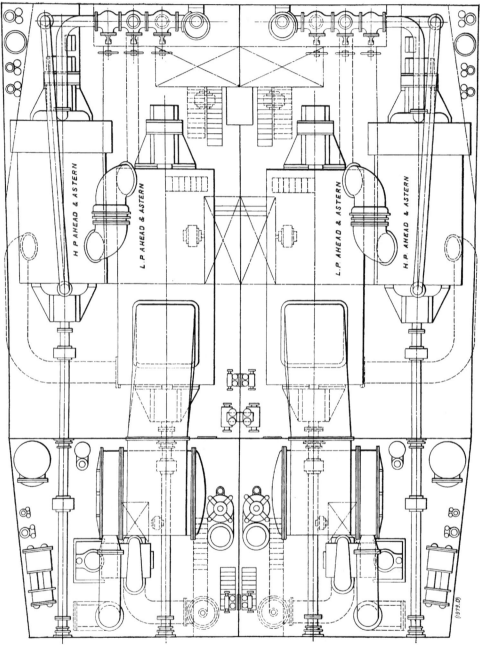

H P AHEAD & ASTERN

L.P AHEAD & ASTERN

L.P AHEAD & ASTERN

H.P. AHEAD & ASTERN

(1979.D)

Plate CXII.—Four=Shaft Installation of Turbine Machinery with Impulse Wheel at Initial End of High=Pressure Ahead and Astern Turbine (Fig. 251).

is avoided. Many engineers consider this an advantage, because it is urged that diaphragms require to be very strong to prevent distortion, and to have a very small clearance round the shaft. If the diaphragm or diaphragms be divided in halves, so as to enable the rotor to be lifted out clear of them, not only is it necessary to increase their scantlings in order to compensate for this weakening, but it is also necessary that no part of a diaphragm should overlap a disc—a condition which further increases the length of the rotor shaft between bearings. Further, the diameter of a disc-bearing shaft being necessarily much smaller than that of a drum, it is much more liable to be deflected; and the slightest eccentricity or vibration from any cause whatever may bring the shaft into contact where it passes through the diaphragms. As experience has proved that almost all troubles with turbines having more than one impulse stage have hitherto arisen either from undesirable features in the construction connected with diaphragms, or from the overloading with multiple discs of a relatively thin shaft, there is inducement to adhere to the drum construction from end to end of a turbine and to use only one impulse stage, which can very well be mounted either on the drum itself or on a disc projecting from it, as shown in the section on Plate CXI.

Naturally, reversing turbines follow the same construction as those for ahead running, with the necessary changes in length and increase in steam velocity.

The turbines above described are, of course, suitable not only for the driving of twin screws, but may be and are also used for the high-pressure turbines of four-shaft arrangements where there is a high-pressure and a low-pressure turbine on each side of the ship as shown in Plate CXII., facing this page. This plan is applicable for all battleships and armoured cruisers and for small cruisers of high power.

A corresponding arrangement of turbines of the impulse-reaction type may also be applied to small cruisers with one engine room abaft the other, the shafts on one side being driven by the turbines in the forward engine room and on the other side by the turbines in the after engine room. Such a disposition possesses all the characteristics of a twin-screw installation, and the subdivision by transverse bulkheads has very great advantages from the point of view of stability. This arrangement is shown on Plate CXIII., facing this page.

In all of these Parsons impulse-reaction combination turbines it is possible to combine the high-pressure reversing turbine with the ahead turbine in one casing, each, of course, exhausting separately to its corresponding low-pressure unit. Such a construction is being carried out at the present time for battleships, as shown on Plate CXI., facing page 154.

With the high-pressure turbine having an impulse initial stage and reaction blading for the remaining stages, there may be no gain in efficiency at full power, but on the other hand, at small fractions of power, the steam consumption will be generally intermediate between that of the arrangements having and not having cruising turbines.

Other combinations are possible where, without increasing the number of turbine units in a ship, an additional impulse velocity-compounded cruising element may be embodied in the casing of a low-pressure main turbine, giving an economy approaching that attained with independent cruising turbines.

The application of the impulse-reaction turbines to a destroyer would follow the same practice as with ordinary reaction turbines arranged in one engine room. On Plate CXIV., adjoining this page, there is illustrated the arrangement of independent impulse-reaction turbines with two engine rooms, one abaft the other, in a twin-screw torpedo-boat destroyer. On Plate CXV., adjoining page 157, there is illustrated a two-shaft arrangement of turbines, each turbine being complete as an independent unit with an impulse wheel at the high-pressure end of the ahead and astern turbines. Any of these arrangements is applicable to small cruisers, where the power is not so high as to render the units too large and heavy for twin screws. The constructions and arrangements above referred to are now coming into use in warships in several navies. Such turbines may also be used for three-shaft vessels with independent turbines on each shaft. Naturally such turbines cannot be expected to attain

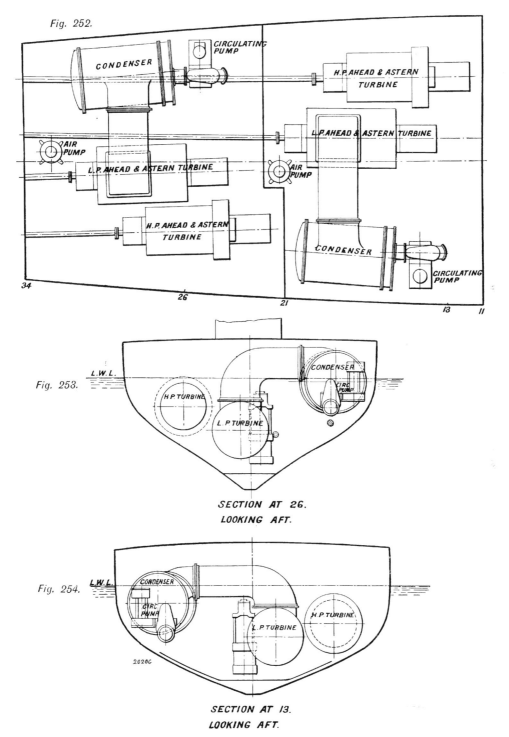

Fig. 252.

CIRCULATING PUMP

CONDENSER

H.P. AHEAD & ASTERN TURBINE

AIR PUMP

L.P. AHEAD & ASTERN TURBINE

L.P. AHEAD & ASTERN TURBINE

AIR PUMP

H.P. AHEAD & ASTERN TURBINE

CONDENSER

CIRCULATING PUMP

34   26   21   13   11

Fig. 253.

L.W.L.

CONDENSER

CIRC PUMP

H.P. TURBINE

L.P. TURBINE

SECTION AT 26.
LOOKING AFT.

Fig. 254.

L.W.L.

CONDENSER

CIRC PUMP

L.P. TURBINE

H.P. TURBINE

2020C

SECTION AT 13.
LOOKING AFT.

Plate CXIII.—Four=Shaft Arrangement in Two Engine Rooms (one abaft the other) of Turbines in Series with Impulse Stage on High=Pressure Ahead and Astern Turbines (Figs. 252 to 254).

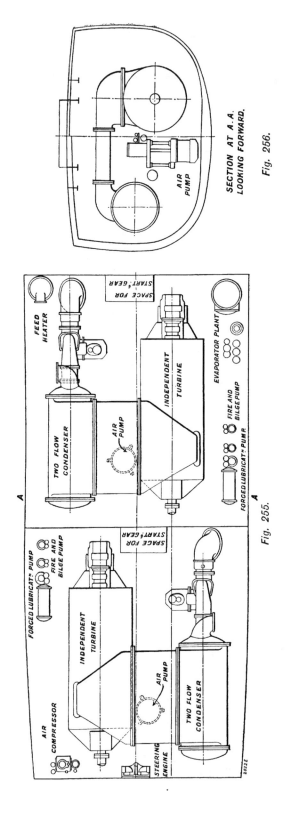

SECTION AT A.A.
LOOKING FORWARD.

Fig. 256.

Fig. 255.

Plate CXIV.—Two=Shaft Arrangement with Independent Parsons Impulse=Reaction Turbines, in Two Engine
Rooms, for Twin=Screw Torpedo=Boat Destroyers (Figs. 255 and 256).

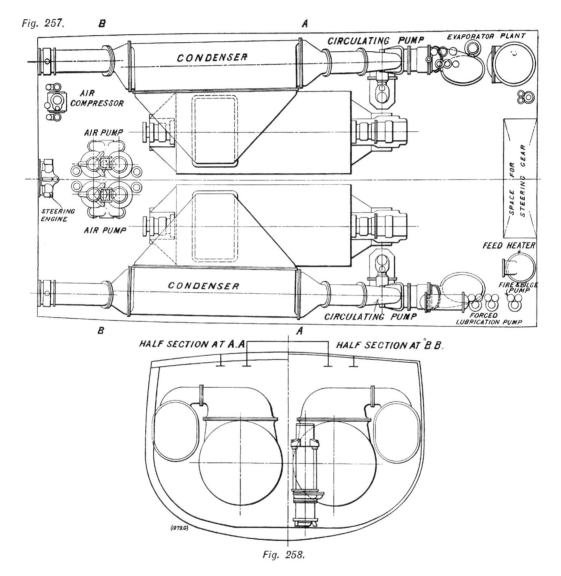

Fig. 257.

B   A

EVAPORATOR PLANT

CIRCULATING PUMP

CONDENSER

AIR
COMPRESSOR

AIR PUMP

SPACE FOR STEERING GEAR

STEERING
ENGINE

AIR PUMP

CONDENSER

FEED HEATER

CIRCULATING PUMP

FIRE & BILGE PUMP

FORCED
LUBRICATION PUMP

B   A

HALF SECTION AT A.A   HALF SECTION AT B.B.

Fig. 258.

Plate CXV.—Two=Shaft Arrangement for Twin=Screw Torpedo=Boat Destroyer, with One Engine Room, the Impulse=Reaction Ahead and Astern Turbines on each Shaft being an Independent Unit (Figs. 257 and 258).

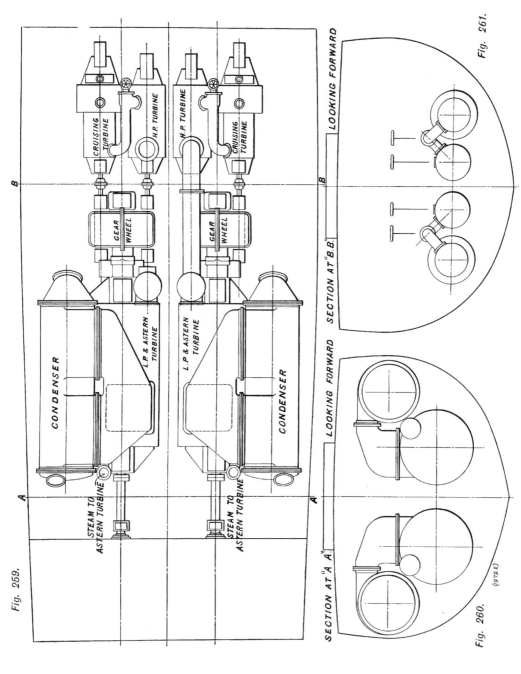

Fig. 259.

STEAM TO ASTERN TURBINE

CONDENSER

L.P. & ASTERN TURBINE

GEAR WHEEL

CRUISING TURBINE

H.P. TURBINE

H.P. TURBINE

CRUISING TURBINE

GEAR WHEEL

L.P. & ASTERN TURBINE

CONDENSER

STEAM TO ASTERN TURBINE

LOOKING FORWARD

SECTION AT "A A"

Fig. 260.

(979.E)

LOOKING FORWARD

SECTION AT "B B"

Fig. 261.

Plate CXVI.—General Arrangement of Geared Cruising and High=Pressure Turbines for Torpedo=Boat Destroyers (Figs. 259 to 261).

quite as small consumption at low powers as the earlier arrangements with series working and cruising turbines.

There are also great possibilities in the interposition of gearing in high-speed warships between the high-pressure and low-pressure turbines, so that each may work at its highest efficiency. The arrangement most favoured at the present time in two-shaft installations is to have a low-pressure turbine and an astern turbine direct coupled to the propeller shaft, and a high-pressure turbine connected at the forward end to the same shaft through helical gearing. By this means the high-pressure turbines may be designed to run at speeds conducive to economy, being designed for the maximum blade efficiency, and at the same time a high initial steam pressure can be maintained, leading to increased economy at all powers, and to high economy at low powers, especially when suitable cruising stages are fitted. The high-pressure turbines may consist of two turbines in series, both of which would be in operation at full power, the higher-pressure turbines having a by-passed cruising stage for use at low powers, or one of the high-pressure geared turbines may be a cruising turbine, solely for use at low speeds and connected to run in vacuum at high speeds. In installations with more than two shafts, the arrangement favoured is one in which small high-speed cruising turbines are geared to the low-pressure shafts, the full power turbines being direct coupled. These geared turbines, it will be seen, simply take the place, in such an arrangement, of the ordinary cruising turbines already alluded to. An arrangement of such geared and high-pressure turbines suitable for torpedo-boat destroyers is illustrated on Plate CXVI., facing this page.

The application of gearing to torpedo-boat destroyers was preceded by extensive research work, and by the successful trials of the "Vespasian," a cargo steamer fitted with turbines driving through such gearing a single screw propeller. This improvement, like all others of the Parsons Company, eventuated from experiments, which are fully dealt with in the succeeding chapter on the developments in the machinery arrangements for moderate-speed and low-speed merchant vessels.

There is further the important question of the economy resulting from utilising in the low-pressure turbine the remaining energy in the steam exhausted from the extensive auxiliary machinery, which absorbs a large proportion of the steam generated, especially when the ship is running at reduced speed. The diagram, Fig. 262, on this page, indicates the steam consumption of the machinery of a recently completed warship at various powers, showing the rate (1) for turbines only per shaft horse-power per hour for all speeds and for all purposes, (2) with open exhaust to the condenser, and (3) with closed exhaust, *i.e.*, when the steam passes from the auxiliaries to the low-pressure turbines. It will be seen that the practical effect is that for a large range of power the steam consumption of the auxiliaries is reduced by half so far as the demand on the boilers is concerned. While some such economy is realisable from the same utilisation of the auxiliary exhaust steam in the case of reciprocating machinery instal-

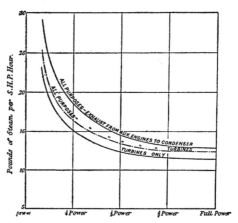

Fig. 262.   Diagram Illustrating Economy of Closed Exhaust.

lations, it is not of anything like the same extent, the turbine system, as already stated, being specially suited to work with low-pressure steam.

Arising out of the subjects discussed in this chapter is the question of parallel or series working. The choice of the number of shafts is becoming more and more limited owing to the great power required to attain the high speed demanded in warships and to the limitations of transmission of such powers through each shaft. It is obvious that the installation for a given power should not be divided into too many independent high-pressure units, as the difficulty of maintaining high initial pressure at low fractions of power is very great. If, for instance, a destroyer having three shafts with one high-pressure turbine and two low-pressure turbines (the whole forming one steam unit) be compared to a twin-screw vessel having two main independent turbines in parallel, the total quantity of steam in the latter case has to be divided over two high-pressure admissions to the turbine instead of one. Not only does this at once halve the height of reaction blading, but, as the revolutions are reduced in the ratio of the square roots of the number of shafts, i.e., $\sqrt{3} : \sqrt{2}$, or $18\frac{1}{2}$ per cent., the diameter of the turbine must be correspondingly increased if the blade speed is to be maintained, and this, of course, reduces the blade height correspondingly in order to keep the area of blade annulus the same. The final result is that the initial blade height in a twin-screw ship with purely reaction blading would only be about 40 per cent. of the height for the single high-pressure turbine of a triple-screw vessel with high-pressure and two low-pressure turbines. This reduced blade height being on a greater diameter, the leakage and clearance losses would become prohibitive.

Where more than two shafts are imperative, owing to the greatness of the power, working in series has thus advantages whether the turbines are of the reaction or combined impulse and reaction type. This series working constitutes one of Mr. Parsons' patents, and, by arrangement, is now being applied to other installations than those with reaction turbines. This practice is likely to be extended, especially as, owing to facility in manœuvring and the subdivision of the ship, four shafts are usually preferred to three. Moreover, the turbines are of less diameter, and can be more easily fitted under the protective deck. With turbines in series, too, the engine room is shorter, and any saving in the length of the ship reduces the cost, especially when the ship is armoured. With four shafts the turbines on each side of the centre line may advantageously be worked in series. In three-shaft arrangements, adopted by some authorities, it is possible to have independent turbines of the impulse-reaction combination type, driving each shaft, though the length of each turbine becomes so great that it is usually advisable to fit tandem turbines on each shaft, making six in all. It cannot be expected that this arrangement will be extensively adopted. For low powers the turbines on the wing shaft may be conveniently run in series with the high-pressure portion of the turbines on the centre shaft, and such an arrangement is being carried out in some vessels now building.

Torpedo-boat destroyers fall into a class by themselves. Simplicity of arrangement is being recognised as a great factor in the machinery of this class of vessel. Several twin-screw vessels are now building for various navies, including the British, German, French, Austrian, Italian, Argentine, Danish, and Swedish. The Parsons impulse-reaction combination turbine lends itself very well to this, and the full power efficiency will probably be nearly equal to that of earlier boats with three shafts and turbines in series, though the consumption at low powers cannot be expected to be quite as favourable as when separate cruising turbines are adopted. Something may, however, justly be sacrificed to obtain simplicity, and, in the case of ships with two engine rooms, a complete independence of each room.

RECENT AND PROSPECTIVE DEVELOPMENTS IN MERCHANT SHIP TURBINES.

THE lines of development in turbines for merchant ships are more clearly defined than in the case of warship work, because there does not exist the same complications due to variable speed as obtain in warship service. The problem set in connection with the design of warship machinery, and discussed in the preceding chapter, is largely associated with the realisation of economy at the lower speeds, at which warships so often cruise, as well as at the higher speeds. Merchant ships, on the other hand, always steam from port to port at or about their maximum speed, and the Parsons reaction turbine working in series on three or four shafts is unrivalled in the efficiency of the results achieved under such conditions. Indeed there are only one or two ships fitted with other types of turbine, so that future developments in the case of high speed merchant ship machinery will probably be largely in details, similar to those reviewed in the chapters describing the turbines of Channel steamers and ocean liners (pages 96 to 147).

It will be seen from our review that hitherto the steam turbine has been adopted for propulsion in vessels having speeds exceeding 17 to 18 knots, and it would appear as if in the future it would completely supersede the piston engine in such steam ships, as the economy and other advantages of the system are becoming universally accepted. But for lower speeds the difficulty of accommodating the comparatively slow speed of revolution which favours efficiency in the screw propeller to the high blade-speed which favours efficiency in the turbine is an obstacle. The rate of revolutions of the propeller has been reduced in recent practice by increasing the diameter of the turbine, so that for lower rates of revolution the peripheral speed of the turbine rotors or blade-speed may continue the same. The limit to the size of the rotor is largely a matter of convenience and weight, but increase in diameter involves a smaller height of blade with the possibility of greater leakage. Gearing may come into use in the future for the combination of a high speed, and, therefore, an economical and light, turbine with a slow speed propeller ; to this reference will be made later. But, for ships of intermediate speeds, Mr. Parsons has introduced a combination of high-pressure reciprocating engines and of low-pressure turbines, which latter ensure the high economy due to great range of expansion and high vacua.

The low-pressure condensing turbine was first designed in 1889 to take the steam exhausted from a high-pressure turbine at atmospheric pressure and expand it down to 1 lb. absolute, and a section showing this machine, as designed in 1889, is given on Plate CXVII., facing page 163. This machine was not made until 1899 for reasons given on page 34. The very high efficiency of the low-pressure turbine, always anticipated, was disclosed by the Elberfeld results (page 64). But prior to this—in 1894—a patent[1] had been taken out for a combination of a reciprocating engine with a steam turbine, in order to increase the power obtainable by the expansion of steam beyond the limits possible with reciprocating engines. In the case of H.M.S. "Velox," built in 1902 (see page 82 *ante*), the reciprocating engines, used only for driving the ship at low power, exhausted into the main turbines, but it was not until 1906 that the idea of a combination was carried out in its entirety. The delay is accounted for by the extreme pressure of business in the development of the turbine for high-speed duty on land and on board ship.

The idea is based on a sound principle, as experience has now proved. There was nothing problematic about the issue. The value of high vacua, and the practicability of efficiently using low-pressure steam, had been proved in connection with the introduction of exhaust turbo-generators on land. (See page 58.) With steam pressures of 15 lb. to 16 lb. absolute per square inch, and a vacuum of 28 in., it was found possible to generate 35 kilowatts of electricity for every 1000 lb. of steam used in the low-pressure turbine.

On Plate CXVII. there is set out in diagrammatic form the advantage of the combined system. The total area of the diagram

[1] Patent No. 367, A.D. 1894.

(Fig. 266) represents the maximum power that could be obtained, theoretically, from the steam if it were expanded down to the pressure of the condenser. The area enclosed by the lines A, B, C, D, and E shows the theoretical maximum energy realisable in a quadruple engine working from 200 lb. pressure to 26 in. vacuum, and the area cross hatched the additional energy that can be utilised in a turbine, but which cannot be economically used in a reciprocating engine. On the same Plate there are two diagrams (Figs. 264 and 265) which show the effect of increased vacuum on the output of exhaust-steam turbines—one using 10,000 lb. of steam per hour at 15 lb. absolute pressure and the other using 30,000 lb. at 175 lb. pressure. In the former case (Fig. 264) the net output with 28 in. vacuum is about 280 kilowatts, and at 29 in. vacuum 335 kilowatts, a gain of about 20 per cent.; while in the case of the turbine working with an initial pressure of 175 lb. per square inch (Fig. 265) the gain due to the same additional inch of vacuum is from 1520 to 1600 kilowatts, or about 5 per cent., a result which shows the great advantage due to high vacua with the low-pressure exhaust turbine.

As far back as 1901 various designs for a combination of reciprocating engines and turbines suitable for moderate-speed boats were prepared by The Parsons Marine Steam Turbine Company, Limited, in association with one or two shipbuilders, and during the years 1907 and 1908 designs of vessels of intermediate power and speed, carrying passengers and cargo, were got out by the Parsons Company in conjunction with Messrs. Swan, Hunter, and Wigham Richardson, Limited, for the application of the combination system.

The New Zealand Shipping Company, who were early in the adoption of the turbine for high speed vessels, were the first to adopt the combination system. This also was due to the enterprise of Messrs. Denny, of Dumbarton. The system was applied in a 14½-knot steamer named the "Otaki," built and engined by them in 1908, and the results, to be dealt with presently, induced the owners to repeat the order in 1909. The White Star Line, when they ordered from Messrs. Harland and Wolff, Belfast, two 16-knot steamers, of 20,000 tons displacement under service condition, for their Canadian Service—the "Laurentic" and the "Megantic"—

decided to fit the former with the system, and a consequence of experience by Messrs. Harland and Wolff was the adoption of the combination machinery in the two largest ships yet launched for the New York trade—the White Star Liners " Olympic " and " Titanic," built in 1909-11.

The same firm is also fitting the combination system in the " Demosthenes," of 19,500 tons displacement, to steam 13 knots, being built for Messrs. George Thompson and Co., Limited, of Aberdeen Clipper fame, and in a large ship for the Royal Mail Steam Packet Company's service to the South American republics.

The Orient Line ordered combination machinery for their new steamer " Orama," laid down in 1910 for their Australian mail service. The vessel is 550 ft. long, 64 ft. beam, and 34 ft. 3 in. depth moulded, and of about 18,000 tons displacement, and is being constructed and engined by Messrs. John Brown and Co., Limited, Clydebank.

All of these ships have triple screws, each reciprocating engine driving a wing shaft and propeller, while the turbine, taking the exhaust steam from both engines, works the centre screw. This arrangement has the advantage that, in manœuvring and for going astern, the piston engines only are used, and thus there is no need for an astern turbine. In the reciprocating engines the steam is expanded from boiler pressure to about 15 lb. to 10 lb. absolute, and passes, in normal working, through a " change " valve to the turbine, where it is expanded to about 1 lb. absolute, exhausting through the usual eduction pipe to the condensers. The interposition of the change valve enables the steam to be passed to the condenser direct in the event of the turbine being disabled, or when the turbine is to be " cut-out " in order that the ship may be manœuvred or driven astern by the piston engines only. In the last contingency the turbine is disconnected from the condensers by the closing of a sluice valve in the eduction pipe. The change valve is, as a rule, operated by the engine which reverses the piston machinery. This " interlocking " arrangement obviates any possibility of the change valve being open to the turbine when the main engines are reversed.

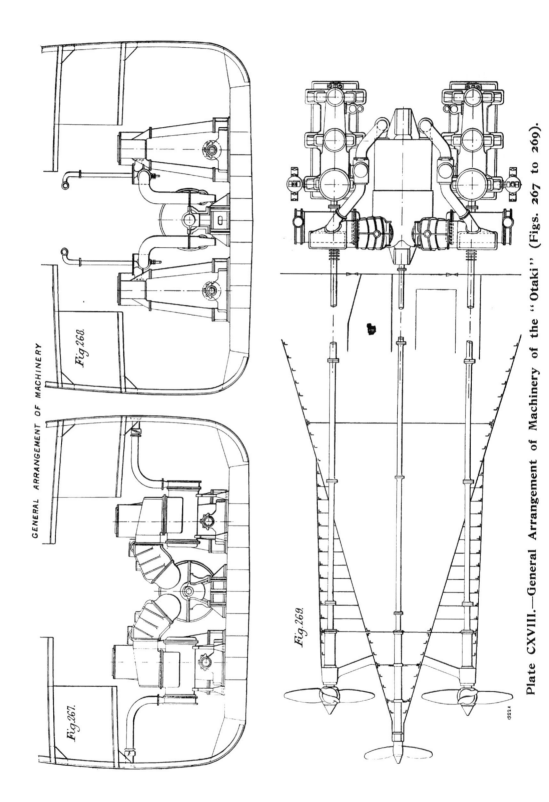

GENERAL ARRANGEMENT OF MACHINERY

Fig. 268.

Fig. 267.

Fig. 269.

Plate CXVIII.—General Arrangement of Machinery of the "Otaki" (Figs. 267 to 269).

Arrangements are also made so that, should one of the piston engines break down, the other engine and the turbine can be used together for driving the ship by means of the propellers on one wing and the centre shafts, the manœuvring and astern work being done by the one piston engine in action. There is gain in the fitting of a turbine in such case, as it can take steam from the boilers and be very considerably over-loaded, increasing the available power up to two-thirds the total power for all three sets. Even in such case there need not be any great falling off in economy. Similarly one of the two condensers can be disconnected in the event of it breaking down or of any of the condenser auxiliary machinery being out of action. To obviate any foreign matter—oil, &c.— passing over with the steam, there is a separator or strainer at the main stop valves and at the change valve.

The "Otaki," the first vessel with the system, is more or less historical, and, as it affords a satisfactory example of the principle and its application, may be fully described. The results of the trials and of those of two sister ships of corresponding dimensions, built contemporaneously and differing only in having twin-screw reciprocating engines, were analysed in a paper[1] read in August, 1909, at the Institution of Engineers and Shipbuilders in Scotland, by Engineer-Commander W. McK. Wisnom, R.N., whose experience of turbine work at the Admiralty, and later as Engineering manager and partner of Messrs. Denny, justifies confidence in the conclusions. The data, he states, are sufficient to show that "a high degree of economy may be anticipated in the combination type of engine—an economy probably higher than has been obtained in any other type of marine engine for vessels of this class. There appears little doubt that in vessels of low speeds, for which Parsons turbines alone are unsuitable, and where economy is of primary importance, this system can be advantageously employed. . . . The possible difficulty in manœuvring has been advanced as an objection to the combination system. No difficulty of any kind in connection with manœuvring was experienced in the 'Otaki.' The maximum power cannot be developed when

[1] See "Transactions of the Institution of Engineers and Shipbuilders in Scotland," vol. lii., page 279 ; "ENGINEERING," vol. lxxxviii., page 179.

going astern, but it was found that astern-power sufficient for all practical purpose had been provided." The general arrangement of the machinery is shown on Plate CXVIII., facing page 165.

The "Otaki" is 464 ft. 6 in. in length between perpendiculars, 60 ft. in breadth, and 60 ft. in depth, and has a deadweight capacity of 9900 tons on a draught of 27 ft. 6 in. She is 4 ft. 6 in. longer than the two sister ships "Orari" and "Opawa," fitted with reciprocating engines, in order to make up for the loss in cargo capacity due to there being three shaft tunnels instead of two. Such extra load put on the "Otaki's" machinery for a given speed is debited to her steam consumption. But experience showed that the reduction in steam consumption not only liberated some part of the area of coal bunkers and some part of the load of coal, but that the boilers and their compartments could have been reduced, making an addition to cargo capacity much greater than the space deducted for the extra shaft tunnel. As it was, the machinery weighed 30 tons, or 3.25 per cent. more than that of the "Orari" and "Opawa."

The steam-generating plant in the three ships is alike—five single-ended cylindrical boilers, with Howden's system of draught, having 305 square feet of grate area, 13,500 square feet of heating surface, and working at 200 lb. pressure. The engine-room is of the same area, and in each case accessibility to all parts was a feature.

The reciprocating engines in the "Orari" and "Opawa" have cylinders $24\frac{1}{2}$ in., $41\frac{1}{4}$ in., and 69 in. in diameter respectively with a stroke of 4 ft. The ratio of high-pressure to low-pressure cylinder is thus 1 : 7.9. In the case of the "Otaki" the ratio is 1 : 5.6; the cylinders being $24\frac{1}{2}$ in., 39 in., and 58 in. in diameter respectively, and both engines exhaust into a turbine of the Parsons low-pressure type. The diameter of the rotor drum is 7 ft. 6 in., and the lengths of blades range from $4\frac{3}{4}$ in. in the first, to $12\frac{11}{16}$ in. in the last, expansion. The drum is completely closed at both ends, and any leakage past the dummy is led away by an external pipe to the condenser. The object of this is to reduce the loss due to the cooling effect of the condenser on the internal surface of the drum, and to prevent corrosion. The arrangement of change valves and gear is shown in Figs. 270 and 271, on Plate CXIX., facing this page.

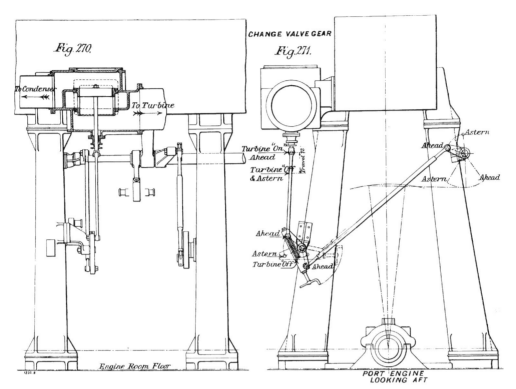

Figs. 270 and 271. Change-Valve Gear.

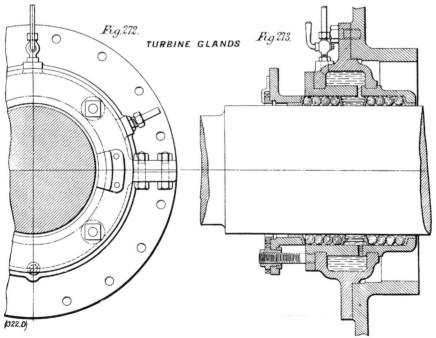

Figs. 272 and 273. Turbine Glands.

Plate CXIX.—Details of the Combination Machinery of the "Otaki."

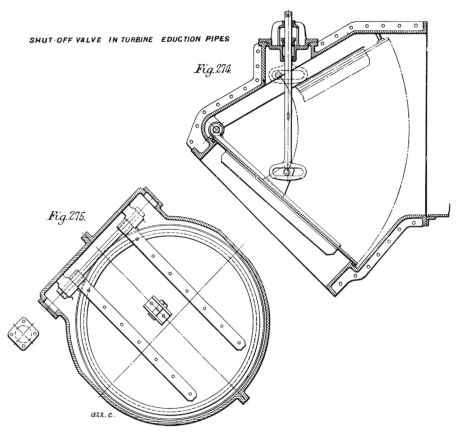

SHUT-OFF VALVE IN TURBINE EDUCTION PIPES

Fig. 274.

Fig. 275.

1322. C.

Figs. 274 and 275. Shut-Off Valve in Turbine Eduction Pipe.

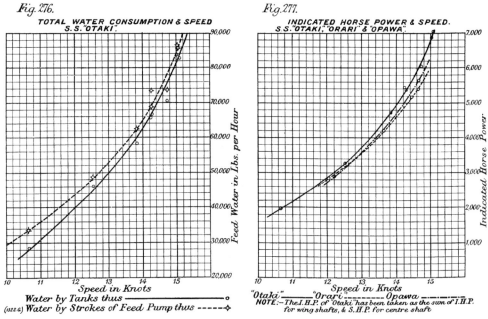

Fig. 276.

TOTAL WATER CONSUMPTION & SPEED
S.S."OTAKI".

Feed Water in Lbs. per Hour

Speed in Knots

Water by Tanks thus ——————o
(1322 G) Water by Strokes of Feed Pump thus -----⟡

Fig. 277.

INDICATED HORSE POWER & SPEED.
S.S."OTAKI", "ORARI" & "OPAWA".

Indicated Horse Power

Speed in Knots

"Otaki"————— "Orari"————— Opawa—·—·—·
NOTE:—The I.H.P. of "Otaki" has been taken as the sum of I.H.P.
for wing shafts, & S.H.P. for centre shaft

Figs. 276 and 277.   Diagrams of Results of "Otaki's" Trials.

Plate CXX.—Details of the Combination Machinery of the "Otaki."

The turbine glands, illustrated by Figs. 272 and 273, on Plate CXIX., facing page 166, are packed with soft material, separated into two parts by a metallic lantern ring. This ring enables the shaft to be surrounded by water between the two divisions of the packing, thus forming a water seal. The water is supplied under a slight head, and a gauge glass at each stuffing box indicates the head of water, and whether air leakage is occurring. This method of packing worked very satisfactorily on service.

The shut-off valve fitted in the exhaust from the turbine to each condenser, to disconnect either condenser, is shown on Plate CXX. These valves have the further utility that they enable the turbine to be cut out for removing any deposit of oil while at sea. But this need, anticipated by some, has not arisen. A large pocket was provided at the lower end of the steam inlet pipes to the turbine with baffle and examination and cleaning door; but, from experience on service, there appears, says Mr. Wisnom, "no reason to anticipate any trouble from this cause."

On the full-power trial on the measured mile at Skelmorlie, the "Orari," with the twin-screw reciprocating engines, obtained a mean speed of 14.6 knots, and the "Otaki," with combination engines,

TABLE X.—SUMMARY OF COMPARISONS OF TRIAL RESULTS OF THE "OTAKI" AND "ORARI" FOR A SPEED OF 14.6 KNOTS ON THE MEASURED MILE.

Name.	E.H.P.	I.H.P.	Propulsive Coefficient.	Water Consumption.		
				Total per hour.*	Per E.H.P. per hour.	Per I.H.P. of "Orari" per hour.*
				lb.	lb.	lb.
"Otaki" ...	3350	5880	57 per cent.	73,300	21.9	13.7
"Orari" ...	3210	5360	60 per cent.	88,300	27.5	16.5
Gain per cent. in "Otaki"				17	20	17

NOTE.—Columns marked * do not take into account the differences of effective horse-power in the two ships; these two methods of comparison should show the same gain.

a mean speed of 15.09 knots. On corresponding basis and for the same speed the water consumption per hour was 88,300 lb. and 73,300 lb. It is admitted by Mr. Wisnom that the records on the measured mile runs only are liable to error on account of the short duration of the observations. It has therefore been considered preferable to plot the results of various observations, to draw curves of mean results, as shown on Plate CXX. and to use figures taken from the curve in making comparison. Care was taken in drawing the curves not to give unduly favourable results.

From a design point of view the most reliable comparison appears to be one based on water consumption per effective horse-power where this is available, while from a practical point of view it is obvious that in sister vessels, with similar boiler installations, the coal consumption per mile or per hour at the same speed affords the best comparison of the two systems.

Table XI. shows the initial pressures, range of temperature, &c., in the different cylinders as taken from two sets of indicator diagrams

TABLE XI.—INITIAL PRESSURES AND RANGE OF TEMPERATURE IN CYLINDERS OF THE "OTAKI" AT FULL SPEED.

	Reciprocating Engines.			Turbine.		
—	H.P. Initial Pressure Diagram.	M.P. Initial Pressure Diagram.	L.P. Initial Pressure Diagram.	L.P. Exhaust Pressure Diagram.	Initial Pressure (Mercury Column.)	Exhaust Pressure (Mercury Column.)
Absolute pressure in pounds	196	91.5	36	11.5	9.8	1
Corresponding temperatures, assuming saturated steam, deg. Fahr. (approximately)	380	321	261	200	192	102
Temperature range, degrees	59	60	61	8 (Drop)	90	
Theoretical available heat units per pound of steam.	200 British thermal units (approximately).			120 B. Th. U. (approximately).		

at full speed in the "Otaki." It will be observed that the range of
temperature in the low-pressure cylinders of the reciprocating engines
is about the same as in the other cylinders ; and that the theoretical
heat available in the turbine is over one-third of the total available
heat.   On service the saving in fuel was 12 per cent.   This latter
gain represents about 750 tons of coal for the round voyage.   As
the ship leaves England with sufficient coal for the outward run, the
reduction in coal supply would add 375 tons to the possible cargo
load.   But, as we have said, the best indication of the satisfaction
of the owners is the repeat order given as a consequence of
experience with the system in service.

The "Laurentic" and the "Megantic," built for the Canadian
service by Messrs. Harland and Wolff, contemporaneously with the
"Otaki," differ only in their machinery, the latter having quadruple
expansion engines, while the former is the first Atlantic liner fitted
with a combination of reciprocating and turbine machinery.[1]   Their
length is 565 ft. 6 in. over all, beam 67 ft. 3 in., and depth,
moulded, 45 ft. 6 in., the registered tonnage being about 15,000 tons,
and the displacement at service draught about 20,000 tons.   They
each carry 260 first-class, 430 second, and 1000 third-class passengers.   There are in both the "Megantic" and the "Laurentic"
six double-ended steam generators for a working pressure of 215 lb.
The proportions of the boilers are the same in each ship, the allowance in heating surface being, in accordance with the White Star
practice, about $2\frac{1}{2}$ square feet per horse-power, while the ratio of
heating surface to grate area is 37 or 38 to 1.   The machinery
of the "Megantic;" the twin-screw ship, is of the quadruple-expansion
type, the cylinders being 29 in., $41\frac{1}{2}$ in., 61 in. and 87 in. in diameter
respectively, with a stroke of 60 in.   In the "Laurentic" the
reciprocating engines are of the triple-expansion type, with four
cylinders, the diameters being 30 in., 46 in., 53 in., and 53 in.,
respectively, with a stroke of 54 in.   The low-pressure Parsons
turbine is in the centre of the ship and abaft the main engines.
The main engines run at about 83 revolutions, while the turbine
makes 220 revolutions.

[1] See "ENGINEERING," vol. lxxxvii., page 598.

z

The turbine, as already indicated, accords with Parsons latest practice, and gives 28 shaft horse-power per ton. The thrust bearing is at the forward end, with the forward main bearings between this thrust and the rotor; the after bearing is abaft the rotor. The blading is arranged for six expansions, the lengths ranging from 6 in. to about 10 in., while the average clearance is about one-tenth of an inch. There is the usual arrangement for adjustments, and the bearings are lubricated by gravity with two stand-by oil-pumps. The shaft bearings have also water circulating round them.

On the official steam trials the water consumption for the main engines was about 12 lb. per horse-power per hour, while in the Atlantic liners with triple-expansion engines the rate is rarely under 16 lb. On service the consumption is 14 per cent. less than in the "Megantic," at the same speed; in other words the "Laurentic" for the same coal steams three-quarters of a mile per hour faster. The financial gain is considerable, for not only is the coal bill reduced, but the increase in the cargo capacity, due to smaller boilers and bunkers being required, is also a consideration in the balance sheet. It is true that the weight of engines and the space occupied by them are greater when the turbine supplements recipro-cating engines; but the reduction in steam consumption justifies less boiler power, and thus the combined weight of the engines, boilers, &c., is not greater.

The trial-trip results showed the possibility of the maintenance of high power with one engine out of service, as the low-pressure turbine developed 4600 shaft horse-power out of a total of 12,400 horse-power, the steam exhausted from the reciprocating engine to the turbine being at about 18 lb. pressure absolute.

The arrangement of the machinery of the White Star liners "Olympic" and "Titanic," the largest vessels yet launched, is of special interest, and the drawings reproduced on Plates CXXI. and CXXII., facing pages 171 and 172, are therefore important. These vessels have a length overall of 882 ft. 9 in., and between perpendiculars of 850 ft., a beam of 92 ft., and at 34 ft. 6 in. draught the displacement tonnage is about 60,000 tons.[1] To give a

---

[1] See "ENGINEERING," vol. xc., pages 564, 620, and 693.

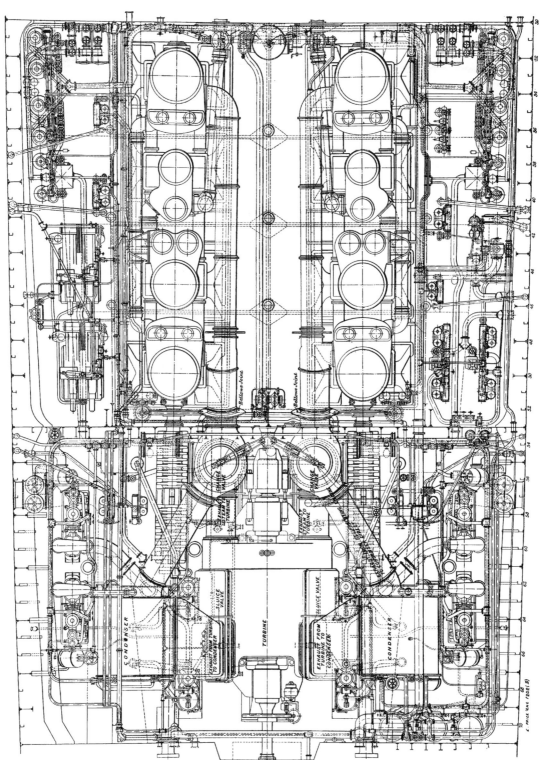

Plate CXXI.—General Arrangement of Combination Machinery in the "Olympic" (Fig. 278).

sea-speed of 21 knots, the reciprocating engines will together indicate 30,000 horse-power at 75 revolutions, and the low-pressure turbine will develop 16,000 shaft horse-power at 165 revolutions.

There are twenty-nine boilers in each ship, having in all 159 furnaces. All of the boilers are 15 ft. 9 in. in diameter, but twenty-four are double-ended, being 20 ft. long, while five are single-ended, being 11 ft. 9 in. long. The working pressure is 215 lb.

A feature of the general arrangement of the combination machinery in this case is that, in view of the size of the units, the exhaust turbine is accommodated in a separate compartment abaft the main reciprocating engine room, and divided from it by a water-tight bulkhead, instead of being in the same engine room with the two sets of piston engines, as in earlier ships. (See Plate CXXI.)

In the reciprocating engine room there are two sets—one driving the port and the other the starboard shaft; in the wings there are various auxiliary engines. In the exhaust turbine room there is fitted, immediately forward of the turbine, the change valves which control the flow of steam either to the turbine or to the condenser. In this compartment there are placed the main condensers, together with their circulating pumps, twin air pumps, &c., evaporators and distilling plant, and pumps for the forced lubrication of all bearings.

The two sets of reciprocating engines are of the four-cylinder triple-compound type, arranged to work at 215 lb. per square inch, and to exhaust at a pressure of about 9 lb. absolute. These engines are of the four-crank type. The high-pressure cylinder is 54 in. in diameter, the intermediate cylinder 84 in., and both the two low-pressure cylinders, 97 in. in diameter, the stroke being 75 in. in all cases. The crank and thrust shafts are 27 in. in diameter, the line shaft $26\frac{1}{4}$ in., and the tail shaft $28\frac{1}{2}$ in., and they are all hollow, the crank and thrust shafts having a 9-in. and the others a 12-in. hole. The propellers have each a cast-steel boss and three bronze blades, the diameter being 23 ft. 6 in.

The exhaust steam turbine has been arranged to expand the steam down to 1 lb. absolute, the condensing plant having been designed to attain a vacuum of $28\frac{1}{2}$ in. (with the barometer at 30 in.), when the temperature of the circulating water is 55 deg. to 60 deg. Fahr.

The rotor is 12 ft. in diameter, and the blades range in length from 18 in. to 25½ in., and are secured on the segmental principle. The length of the rotor between the extreme edges of the first and last ring of blades is 13 ft. 8 in. The casing is of cast iron. The weight of the rotor is about 130 tons, and of the turbine complete 410 tons. The turbine shaft is 20 in. in diameter, the tail shaft 22½ in., each with a 10-in. hole bored through it. The propeller driven by the turbine is solid, of manganese bronze, with four blades, the diameter being 16 ft. 6 in.

The exhaust pipes from the low-pressure cylinders connecting with the change valve are, as in the "Laurentic," fitted with bellows joints, which consist of two flattened conical discs with special steel rings, and with flanges to take the pipes. The conical form of the disc plates enables any difference in length, due to expansion or contraction, to be taken up. In order to ensure absolutely air-tight joints, all the pipes in proximity to the condenser are fitted with these bellows joints. (See Plate CXXI., facing page 171.)

The change valves for shutting off steam to the turbine and simultaneously opening it to the condenser direct, for manœuvring purposes, are of the piston type with a ring of special form. When the pistons of these valves are in their highest position, steam has a clear flow to the strainer and thence to the turbine; when the piston is lowered the connection to the strainer is closed, and that to the condenser is opened. The piston of each change valve is suspended from one end of a suitably-mounted lever, the other end of which is connected to one of Brown's engines of the hydraulic type adopted in reciprocating - engine practice; this engine is actuated from the starting platform in the reciprocating engine room. The mechanism for operating the valve is well shown in Fig. 279, while the valve chamber is shown on Fig. 280, also on Plate CXXII., facing this page.

The new mail steamer for the Orient line, which is to be fitted with the combination system of machinery, is a triple-screw steamer of the same type as, but slightly larger than, the five twin - screw vessels which inaugurated the new Australian mail

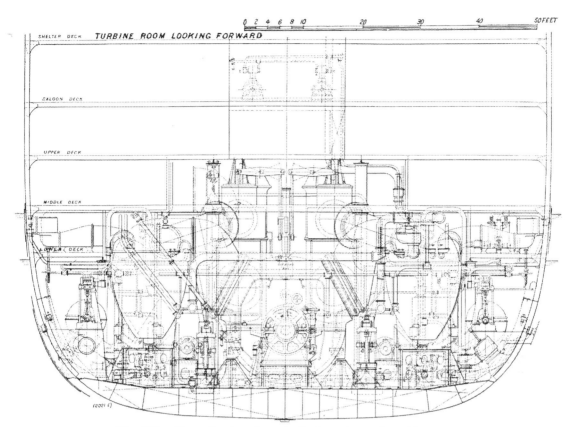

Fig. 279. Cross Section through Turbine Room in "Olympic."

Fig. 280. View of Change Valve.

Plate CXXII.—The Combination Machinery of the "Olympic."

service early in 1910.[1]   The new ship will be of over 15,000 tons displacement on her mean service draught, and will steam on trial over 18 knots.

There are other possible arrangements of combined piston and turbine engines.   The yacht "Emerald" represents an alternative. This vessel, as described on page 104, was originally fitted with a high-pressure turbine on a centre shaft and low-pressure turbines on the wing shafts.   A high-speed enclosed self-lubricating engine, to run at 350 revolutions per minute, was substituted for the high-pressure turbine, and it exhausts at a pressure of about 15 lb. absolute to the two low-pressure turbines.   This has given satisfactory results.

Other variations are possible according to the total number of shafts or the number of reciprocating engines or turbines desired. In association with a single reciprocating engine there could be one turbine for twin-screw working, or two turbines in "series," or in "parallel," with three shafts in all.   With two reciprocating engines there could be, as is usually the case, one turbine in the centre of the vessel, making three shafts in all, or two turbines in "parallel" or in "series" with four shafts.   Reference is made on page 129 to a four-shaft arrangement with two piston and two turbine engines now being constructed in France.   The system which seems to commend itself generally to shipowners and builders, where twin-screw reciprocating engines are fitted, is the arrangement of the turbine on a centre shaft.

Nor is there any restriction regarding the pressure of the exhaust steam from the reciprocating engines, so that were compound engines preferred for their simplicity and for their low initial pressure, which latter lessens the first cost of the boilers, the turbine could be arranged to suit.   From an estimate of the theoretical efficiency, under various conditions of working, it appears, apart from any practical considerations, that there is nothing to choose between 7 lb. and 15 lb. absolute, as any pressure within these limits appears to give economical results.   For a vessel which continually runs at her designed speed an initial pressure of 7 lb. absolute at the turbine appears to be the most suitable ; but for a vessel which runs partly at her designed

[1] See "ENGINEERING," vol. lxxxvii., pages 715, 745, 863 ; vol. lxxxviii., pages 59, 580.

speed and partly at reduced power it is desirable to design the turbines so that the initial pressure shall not fall below 7 lb. absolute when running at moderately reduced proportions of the full power.

It will thus be seen that in addition to its economy the combination system is practicable and adaptable for a great variety of low-speed and intermediate-speed craft, and when one contemplates the millions of tons of coal consumed each year in the world's fleet of merchant steamers, it will be recognised that a reduction of 15 per cent., or even 10 per cent., in fuel consumption must amount to a large saving. Moreover, there is a corresponding reduction in the size of boilers and bunkers and stokehold, with a proportionate increase in the cargo capacity. The system, if widely adopted, must thus yield collectively a handsome profit.

In order to afford to the cargo-carrying steamer of 8 to 10 knots the advantage conferred by the great range of expansion of the turbine and the low consumption of coal resulting therefrom, Mr. Parsons entered upon experiments about 1896 with the view of devising a practical application of a long entertained idea of introducing gearing between the turbines and the propeller. The aim was to enable the turbine to run at a speed which would give maximum efficiency while the propeller continued to revolve at its usual low rate. What is believed to be the first application of turbines driving a propeller through helical spur gearing was made in 1897 by The Parsons Marine Steam Turbine Company, Limited, in the launch built for Mr. F. Buddle Atkinson, and fitted with a geared turbine. This turbine was of 10 horse-power, geared to two wheels, each wheel driving a propeller shaft as shown on Plate CXXIII., facing this page. The revolutions of the propellers were 1400 per minute, and the ratio of gear 14 to 1. The turbine was of the Parsons type with a reversing turbine on the same shaft. Single helical gear was used. The launch was 22 ft. in length and attained a speed of 9 miles per hour and proved very satisfactory. Plate CXXIV., facing page 175, gives a perspective view of this launch at full speed, with details of the turbine and gear.

Exceedingly economical results were also got with a 150 kilowatt

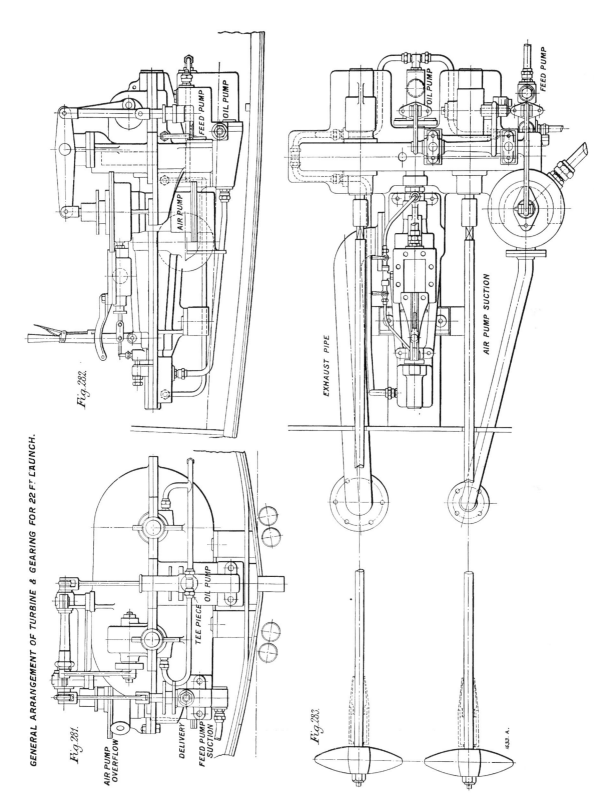

GENERAL ARRANGEMENT OF TURBINE & GEARING FOR 22 FT. LAUNCH.

Fig. 281.

AIR PUMP
OVERFLOW

DELIVERY

FEED PUMP
SUCTION

TEE PIECE

OIL PUMP

Fig. 282.

FEED PUMP

OIL PUMP

AIR PUMP

OIL PUMP

FEED PUMP

EXHAUST PIPE

AIR PUMP SUCTION

Fig. 283.

1633. A.

Plate CXXIII.—Experimental Installation of Turbine and Gearing for 22-ft. Launch (Figs. 281 to 283).

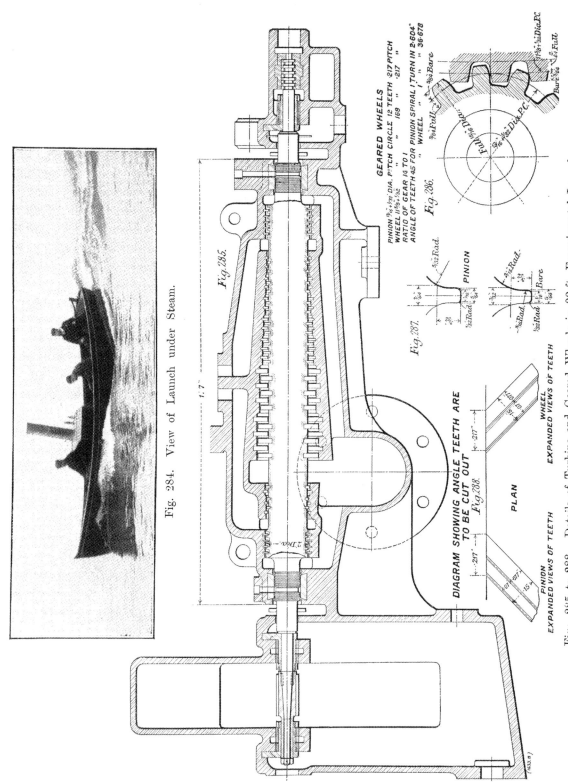

Fig. 284. View of Launch under Steam.

Fig. 285.

1 : 7

GEARED WHEELS

PINION 13/16 + 1/32" DIA. PITCH CIRCLE 12 TEETH · 217 PITCH
WHEEL 11 5/8 + 1/32" " 169 " · 217 "
RATIO OF GEAR 14 TO 1
ANGLE OF TEETH 45° FOR PINION SPIRAL 1 TURN IN 2·604"
WHEEL " 1 " " 36·678

Fig. 286.

Full 9/16 Dia.

9/16 + 1/32 Dia. P.C

1/16 + 1/32 Dia. P.C
1/32 Full.
Bare 9/64
Bare 9/64

Fig. 287.

PINION

PINION
EXPANDED VIEWS OF TEETH

DIAGRAM SHOWING ANGLE TEETH ARE
TO BE CUT OUT

Fig. 288.

PLAN

WHEEL
EXPANDED VIEWS OF TEETH

2 Diam.

Figs. 285 to 288. Details of Turbine and Geared Wheels in 22-ft. Experimental Launch.

Plate CXXIV.—Experimental Installation of Turbine and Gearing for 22-ft. Launch.

turbo-electric alternator supplied to the Newcastle and District Electric Lighting Company in 1898, in which gear was fitted between the turbine and the alternator to reduce the revolutions from 9600 to 4800 per minute. Tests showed a gear transmission efficiency of over 98 per cent.

The Parsons Marine Steam Turbine Company, Limited, purchased in 1909 the steamer "Vespasian," 275 ft. in length and of 4350 tons displacement, in order to fit turbines working the single screw through gearing, in place of the triple-expansion engines with which she was originally fitted. The "Vespasian" had originally been fitted with an ordinary triple-expansion surface-condensing engine by Messrs. G. Clark, Limited, of Sunderland. The cylinders were $22\frac{1}{4}$ in., 35 in., 59 in. in diameter respectively, and of 42-in. stroke. The boilers—two in number—are 13 ft. in diameter by 10 ft. 6 in. long, with a total heating surface of 3430 square feet, and a grate area of 98 square feet, working at a pressure of 150 lb. with natural draught. The propeller is of cast iron, and has four blades, having a diameter of 14 ft., a pitch of 16.35 ft., and an expanded area of 70 square feet.

With a view to obtaining data for comparison between the economy of the original machinery and the turbine with gearing, the reciprocating machinery was first opened out and all such repairs and readjustments were made as were necessary to bring the machinery into an efficient condition and first-class working order. After this had been done the vessel was sent on a voyage, complete data being taken regarding the performance of the reciprocating engine. In order that reliable measurements of water consumption might be made, two tanks were fitted, each of 400 gallons capacity, with suitable change cocks and connections, so that the air pump could discharge through the tanks alternately.

After the completion of this voyage the reciprocating engines were removed and the geared turbines substituted. The only alteration made was in the type of propelling engines; the boilers, propeller, shafting, and thrust block remained the same as for the reciprocating engine. The geared turbine arrangement, as shown on Plate CXXV., facing page 176, consists of two turbines in "series," viz. :—on the starboard side of the vessel one high-pressure turbine, and on the port side one low-pressure turbine, in the exhaust casing of which latter an astern turbine is incorporated. At the after end of each turbine a driving pinion is connected with a flexible coupling between the pinion shaft and the turbine; both pinions are geared into a wheel which is coupled to the propeller shaft. This is well shown in the engraving (Fig. 295) on Plate CXXVI., facing page 177.

The air, circulating, feed, and bilge-pumps are of the usual design for tramp steamers, and are driven by means of a crank and connecting rod coupled to the forward end of the gear-wheel shaft. Forced lubrication is applied to the turbine and pinion-shaft bearings. The teeth of the pinions and of the gear wheel are lubricated by means of a "spray" pipe extending the full width of the face of the wheel.

The high-pressure turbine has a maximum diameter of 3 ft. and an over-all length of 13 ft.; the low-pressure turbine is 3 ft. 10 in. in diameter and 12 ft. 6 in. in length. The turbines are similar in design to those in use in land practice, being balanced for steam thrust only, the propeller thrust being taken up by the originally-fitted thrust block of the horse-shoe type, placed aft of the gear wheel. A new condenser, with a vacuum augmenter, was supplied along with the turbine installation. The cooling surface of the condenser is 1165 square feet.

On Plate CXXVI. there is given a perspective view of the gearing and detailed drawings of the wheel. This gear wheel is of cast iron, with two forged steel rims shrunk on. The diameter of the wheel is 8 ft. $3\frac{1}{2}$ in. pitch circle, there being 398 double helical teeth with a circular pitch of 0.7854 in. The total width of the face of the wheel is 24 in. The inclination of the teeth to the axis is 20. deg. The pinion shafts are of chrome nickel steel, with a pitch circle 5 in. in diameter, with 20 teeth having a circular pitch of 0.7854 in. The ratio of gear is 19.9 to 1.

Comparative results of the water consumption per hour between the geared turbine installation and the reciprocating engine at the several speeds of revolution and under similar

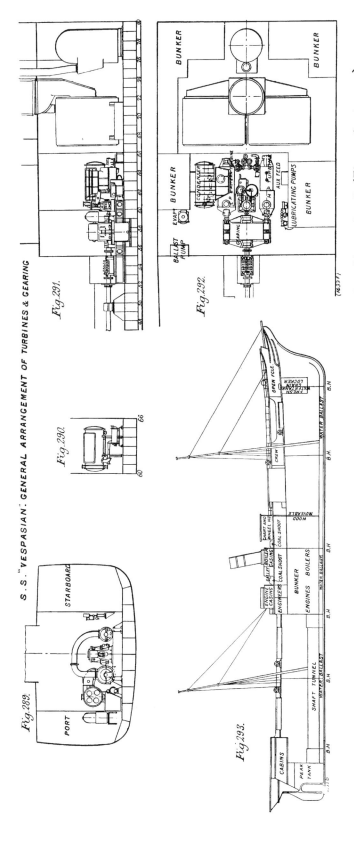

S.S. "VESPASIAN": GENERAL ARRANGEMENT OF TURBINES & GEARING

Fig. 291.

Fig. 292.

BUNKER

BUNKER

CONDENSER

EVAP^R BUNKER

AUX FEED

LUBRICATING PUMPS

BUNKER

BALLAST PUMP

GEARING

Fig. 290.

Fig. 289.

STARBOARD

PORT

Fig. 293.

CABINS

PEAK TANK

B.H

SHAFT TUNNEL

WATER BALLAST

B.H

ENGINE CASING

ENGINEERS

ENGINES

BOILER CASING

BALLAST

COAL SHOOT

BUNKER

COAL SHOOT

BOILERS

B.H

SHAFT AND WHEEL H.ᴰ

WOOD MOVEABLE

CREW

WATER BALLAST

OPEN F'CLE

FRESH WATER TANK

BOSUN LOCKER

B.H

Plate CXXV.—General Arrangement of Geared Turbine Machinery in S.S. "Vespasian" (Figs. 289 to 293).

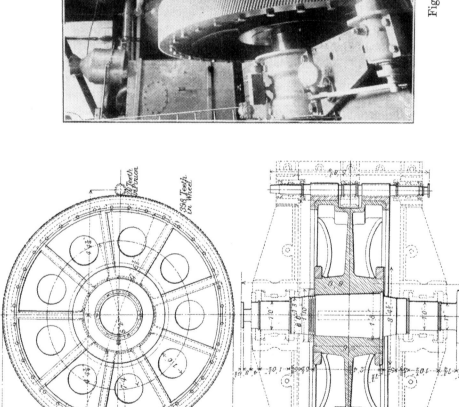

Fig. 295. View of Gear in Ship.

Fig. 294. Details of Gear

Plate CXXVI.—The Gear of *S.S.* "Vespasian."

loaded conditions and draught of water, are given in the diagram, Fig. 296, on this page.   It will be seen that the steam consumption of the turbines at 65 revolutions is 15 per cent. less than that of the original triple expansion engines; at 70 revolutions it is 19 per cent. less.   At 63 revolutions, which gave the continuous economical sea-speed of the vessel when fitted with reciprocating engines, the steam consumption is 13½ per cent. less than with the

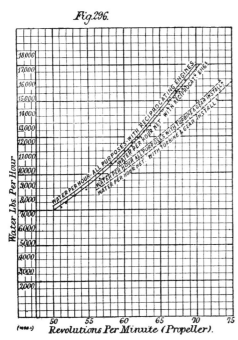

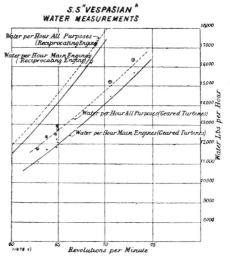

Fig. 296.   Water Consumption per Hour of "Vespasian" with Reciprocating Engines and with Geared Turbines.

Fig. 297.   Steam Consumption of "Vespasian" in Service with Turbines, with Open and Closed Exhaust.

piston engines.   In addition there is a saving in oil and wear, so that the collective gain is considerable.

Since the trials were made with the new machinery, the vessel has been continually in service carrying coal from the Tyne to Rotterdam, and returning in water ballast.   In the year 1910 the vessel steamed over 12,000 miles, and during this period experienced some exceptionally rough weather, especially when steaming in light condition.   The machinery throughout has worked with entire satisfaction, and has not given the least trouble whatever, the most noticeable feature, perhaps, being the remarkable freedom from racing of the engines in rough weather.   Under these con-

2 A

ditions, with the propeller at times entirely out of the water, the acceleration of the engines has never been observed to be above 15 per cent. There is no appreciable wear on the teeth of the gear wheel or pinions. The results of water consumption per hour taken on service are given on the diagram in Fig. 297, on page 177. These results, it will be seen, agree very closely with the trial results referred to previously. The displacement of the vessel on service is about 5 per cent. greater than on the original trials. The results of the working of the "Vespasian" should give shipowners every confidence in the reliability in running and in economy obtainable by this new application of the turbine to marine propulsion.

With regard to the gearing itself, the anticipated high efficiency of mechanical gearing has been fully confirmed. Whilst no actual measurements of this efficiency were made in the "Vespasian," the rise in temperature of the oil supplied to the gearing, which was at first separate from that supplied to the turbines, was so small as to be inappreciable. In experimental gears which were previously tested for this purpose, it was found that the loss in transmission was not more than about $1\frac{1}{2}$ per cent., giving an efficiency of $98\frac{1}{2}$ per cent. It may be taken that a similarly high efficiency of transmission was obtained in the "Vespasian."

Several other systems of power transmission between the turbines and the propeller have been proposed. Electrical authorities advocate the running of turbines at their most economical speed, to generate electricity which would then be utilised in motors rotating the propeller shaft. There might be economy at low powers, as the number of electric units in use could be arranged to suit the power required. The efficiency of the variable-speed motors, if such were used, might diminish at reduced speeds. Moreover, motors that will run at any desired speed are complicated in design and relatively heavy. There are difficulties, too, in respect of ventilating the electric machinery in a confined engine room, while sea-spray, through ventilators, might play havoc with the insulation. This insulating difficulty, however, may be overcome by a system proposed by Mr. Parsons and his colleagues, Messrs. Stoney and Law.[1]

[1] See Patent No. 6177, A.D. 1909.

But it remains to be seen whether the merits of electric transmission of power outweigh the demerits. There is no reduction in space in the proposed schemes; weights in the engine room are increased, and little diminution in boiler weights would be justified. The transmission would involve a loss of about 12 per cent. The conclusion come to by the Engineer-in-Chief of the Navy (Engineer Vice-Admiral Sir Henry Oram, K.C.B.) is that " the interposition of a dynamo and motor between the engine and its work, involving, when on board ship, considerable liability to accident and derangement of the electrical machines, in addition to present risks, has disadvantages such that the practicability of the details should be beyond question, and the gain by its adoption must be shown to be substantial before the system could be fitted to any important vessel."[1] It has also been proposed to utilise hydraulic transmission gear, and one system has been tried on a small scale,[2] that of Dr. Föttinger of Dantzig.

With this reference to the possibilities of the various systems of speed-reduction appliances, we depart from the subject of the application of the marine steam turbine. There is great advantage from the use of the marine turbine in warships and in high-speed merchant vessels, not only in respect of the attainment of speeds otherwise impossible, owing to the higher efficiency of the turbine system, but also because of the economy and the reduction in space and weight attained; but in addition there are very great potentialities for the system in slow-speed craft. The gears we have described make this application thoroughly practicable; but it would seem, especially for cargo ships, that the greater simplicity and higher efficiency of the mechanical gear will commend itself. In such vessels the staff is limited, and repairing appliances are neither numerous nor comprehensive, so that complications in details are to be avoided, and the helical spur gear meets this desideratum more than any other form of gearing yet devised

[1] "Transactions of the Junior Institution of Engineers," vol. xx., page 128.
[2] See " ENGINEERING," vol. lxxxviii., page 601.

ELECTRIC GENERATORS AND TURBINES FOR DRIVING THEM

THE adoption of the turbine for driving the screw propellers of steamships has had first and full consideration in this work, because it appeals more largely to the public imagination. The steamship, like the locomotive, the motor-car, or the flying machine, attracts enormously greater attention because the results are more obvious than is the case with, say, the electric generator or turbo - blower or pump. Thus the lay mind did not fully awaken to the existence of the epoch-marking invention of the Parsons steam turbine until the "Turbinia" made her almost meteoric flight through the Victorian Jubilee Naval Review fleet at Spithead in 1897, although scientists had had their interest quickened during the preceding thirteen years owing to the results achieved by the same prime mover driving electric generators and other machines, and to the important results achieved, and likely to be achieved, by the modifications in the designs of dynamos to meet the new conditions, and particularly the high speed of rotation.

The story of the evolution of a satisfactory electric generator to be driven by the turbine is no less fascinating than the record of experimental work which has perfected the turbine itself; there is, indeed, a close correlation between the two, and, as a matter of fact, Mr. Parsons dealt with them simultaneously. But for the need of a high-speed prime mover to rotate the armature of a dynamo the steam turbine might not so readily have achieved commercial prominence, and similarly the turbo - generator could not have been made efficient without a completely new design of electric machine. This work was of great benefit to electrical science, just as the research on the turbine-driven screw propeller assisted in advancing our knowledge of propellers. Here, however, the analogy ceases. The propeller is most efficient at relatively low speeds of rotation; with the dynamo the reverse is, to a large degree, true. Thus, in the early days, before the advent of a successful turbine, belt or other gearing had to be introduced in many cases between the reciprocating engine and the electric generator to increase the speed of the latter to 1000 or 1500 revolutions per minute from the slower speed of the reciprocating engine. Now, gearing is being applied between the turbine and the screw propeller to reduce the rate of revolution of the latter in the case of slow-speed ships, but for high-speed ships improvements in both turbine and propeller have brought a satisfactory compromise with a high combined efficiency.

The steam turbine required that the speed of the electric generator should be increased, with direct drive, to ten or fifteen times that possible with belt drive from the reciprocating engines of 1884. The first turbo-generator,[1] then introduced, ran at 18,000 revolutions. The armature was $2\frac{5}{8}$ in. in external diameter, so that the surface speed was over 200 ft. per second. The centrifugal force was about 12000 times gravity; in other words every pound weight had a centrifugal force of about $5\frac{1}{2}$ tons. The provision made to counteract such disintegrating force in a built-up armature required the solution of difficult problems. High-carbon steel was used for the spindle. The diameter was $1\frac{1}{4}$ in. at the centre and $\frac{1}{2}$ in. at the bearings, which were $3\frac{1}{2}$ in. long. The core was smooth, $2\frac{1}{4}$ in. in diameter, with notched driving discs of red fibre at each end, and on it was wound a drum winding with barrel ends, the whole being held together by binding wire to withstand the centrifugal force.

Another feature was the spiral end windings and the construction of the commutator segments in short lengths dovetailed into steel rings with asbestos insulation. The magnetic densities, ampere-turns per inch diameter of core, and other electrical quantities, thus early introduced, agree closely with the allowances which would be made to-day under equal conditions. The difficulties were considerable when one recalls the paucity of knowledge then available for designers as regards magnetic densities, hysteresis and eddy currents. It was even doubtful, at that time, whether reversals of magnetism in iron at such rates as 300 per second were possible. The hysteresis losses must then have been four times what they are to-day in view of the more suitable iron now available for core

[1] Patent No. 14723, A.D. 1884.

plates; and because of eddy currents the plates had to be made as thin as possible.

A perspective view of this first high-speed electric generator with the first turbine is given on Fig. 12, Plate X., facing page 25. Sections of the dynamo are given on Figs. 298 and 299, Plate CXXVII., facing this page, while a perspective view of the armature is shown by Fig. 300. This machine was driven by the first steam turbine built, and the consumption, non-condensing, was 129 lb. per kilowatt-hour, with steam of an initial pressure of 80 lb. per square inch. Although only of 8 effective horse-power, this turbo-electric set sufficed to prove the efficiency of the system, and another set was at once made for the electric lighting of the steamer " Earl Percy." A set was also made in 1885 for lighting electrically the Elswick shipbuilding yard. Others followed. The first Atlantic liner lighted with electricity generated by a turbo-dynamo was the " City of Berlin," in which an installation was fitted in 1888. In 1889— five years after the introduction of the system—turbo-generator sets of 60 kilowatts at 6000 revolutions were being manufactured. The dynamos were of the bi-polar type, with shunt-wound horse-shoe magnets. The cores were 7 in. in diameter and 18 in. long, and the commutator segments were usually eighteen in number for 80 or 100 volts. In all but the smallest sizes there was only one complete armature turn per segment.

In the earlier machines a flow of oil passed through a hole in the armature shaft for cooling purposes, but this was found relatively ineffective and was discarded. To minimise sparking, and to avoid the necessity for shifting the brushes, there was introduced compensating winding through holes in the cast-iron pole faces (Fig. 301, on Plate CXXVIII.). This was discarded at that time, as it was thought not to be beneficial, the reason, disclosed by later knowledge, being that its ampere-turns were insufficient to cause a marked effect, being only about three-quarters of those on the armature. This idea was sound, for, as will presently be explained, compensating windings on a similar principle, but with ampere-turns two to two and a-half times those on the armature, proved of great advantage when introduced in 1903.

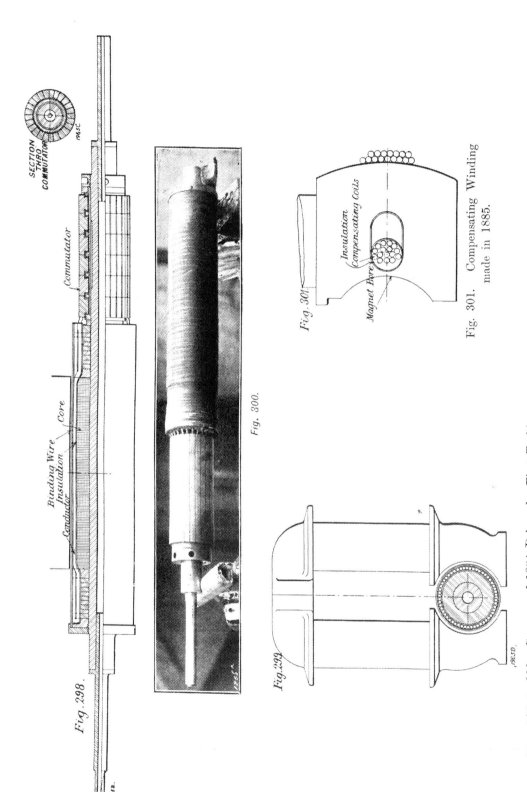

SECTION THRO COMMUTATOR

Commutator

Binding Wire
Insulation
Conductor
Core

Fig. 298.

Fig. 300.

Fig. 299.

Fig. 301.

Insulation
(Compensating Coils
Magnet Bore

Fig. 301.   Compensating Winding
made in 1885.

Figs. 298 to 300.   Dynamo of 1884 Driven by First Turbine.

Plate CXXVII.—Early Turbine-Driven Electric Generators.

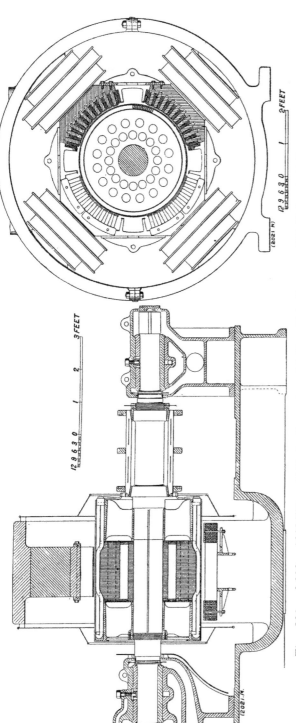

Figs. 302 and 303. 1000-Kilowatt Turbo-Generator with Four-Pole Dynamo, 500 Volts, 1200 Revolutions.

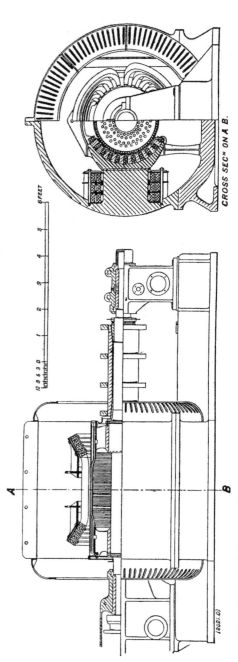

CROSS SEC^N ON A B.

Figs. 304 and 305. 750-Kilowatt Turbo-Generator with Two-Pole Dynamo, 480 to 560 Volts, 1800 Revolutions.

Plate CXXVIII.—Turbo=Generators, illustrating Armatures as still made.

Thus early there were established the following advantages of the turbo-electric generator : steadiness of the electric current produced due to the high speed and the momentum stored in the moving parts, freedom from accident on account of simplicity and direct action, small first cost and maintenance charges for turbine and dynamo, low consumption of oil, reduction in attendance, and minimised size and weight for the power developed — then about 9 watts per pound of weight of the whole machine, including engine and dynamo.

As was the case with the turbine patents, the dynamo inventions for spiral drum-end connections, &c., were retained by Messrs. Clarke, Chapman and Co. when Mr. Parsons ceased to be a partner. Design work had to be commenced *de novo* at the newly-established works at Heaton, in 1890, with the disability that arrangements based on previous experience had to be avoided. There was, however, distinct gain, as new and improved details were evolved, which added later to the efficiency of the machines made according to the dominant idea of the primary patents when they had been re-purchased. To provide against the excessive centrifugal force, the commutators were built up of steel rings shrunk on over mica to hold the segments in place. This construction is now universally adopted by all makers of turbo-dynamos. Fan blades were fitted to the commutator rings to cool both armature and commutator. A drum armature was tried with insulated discs at each end of the core, each disc being connected to opposite conductors. This seemed to give the same advantages as drum-winding with spiral-end connections ; but it was costly and difficult of manufacture. A Gramme armature was therefore adopted for the machines made between 1889 and 1894 ; the most notable were two 410-kilowatt plants having two armatures mounted on the same spindle with their commutators at opposite ends. These, with slight modifications, are still running at the Newburn station of the Newcastle and District Electric Lighting Company. The great trouble with Gramme armatures was the heating of the spindle due to its revolving in the stationary internal field of the armature, and also the impossibility of getting sufficient internal ventilation.

When, however, Mr. Parsons recovered, in 1894, the rights of his earlier patents he reverted to the drum armatures which have since been exclusively used for generators. Among the first improvements adopted was the forming of ventilator ducts along the core for internal cooling, as illustrated in the turbo-generator shown in Figs. 302 and 303, on Plate CXXVIII., facing page 183. The cores were made smooth as the lower self-induction of the windings gave better commutation than with slotted cores. The windings were more thoroughly sub-divided to obviate eddy currents. In some cases the conductors were laid in shallow grooves in the core to prevent them from moving, while the iron between the conductors reduced the reluctance of the air-gap to some extent without perceptibly preventing good commutation. For binding the conductors to the core steel piano wire was preferred to phosphor bronze, because experiments as early as 1887 proved that the loss due to the magnetic properties of steel varied from only 2 per cent. in 12-kilowatt, to 0.30 per cent. in 150-kilowatt generators; while in the case of phosphor bronze the weight and cost were greater owing to the lower tensile strength of the bronze. The tendency also of the bronze to have occasional soft spots of weak material affected its reliability. These are the outstanding characteristics of the armature as still made, and as shown in the engraving on Plate CXXVIII., but some of the latest designs had also radial ventilations, as shown in Figs. 304 and 305.

The commutator is still built up of steel rings shrunk on over mica, and supported on cones at each end. One of the cones is so arranged that it may slide longitudinally along the shaft to allow for expansion or contraction of the copper. In recent machines an expansion diaphragm has been used which obviates any danger of the sliding surfaces sticking, and of the shaft being thrown out of truth by the expansion of the copper segments. In building the commutator, the bars, with mica insulating strips attached, are first assembled and held in place by massive clamps or by one or two layers of steel binding wire put on temporarily. The mica for one of the rings is then built up by slipping pieces of leaf mica under large rubber bands; when a sufficient thickness has been obtained, the mica is tied in place with string and the

rubber bands removed. A steel ring is then slipped on at a red
heat, stops being arranged to make it take up its proper position.
As it goes on it burns away the string and grips the mica so
tightly that the latter can readily be turned and trimmed back
close to the ring, when it presents a smooth black surface like a
solid mica sleeve. In a commutator with only one section there are
two of these steel rings, one at each end; but in long commutators
several are required to counteract the centrifugal force tending to
bend the commutator bars.

The most important improvement introduced was perhaps the
provision made to prevent sparking. It is now well known
that the steam turbine can be very considerably overloaded, but
this could not avail the engineer of an electric-light station, when
confronted with a sudden increase in his load, unless the electric
generator could also be overloaded with equal safety. In his earlier
dynamos Mr. Parsons discarded compensating windings as ineffective,
and in 1900, in association with Mr. Stoney, introduced an invention[1]
for automatically shifting the brushes on the commutator by moving
the brush rocker through a piston controlled by a spring and actuated
by the steam pressure at the inlet to the turbine. The required
angular movement of the brushes as well as the steam pressure at
the inlet are approximately proportionate to the load, but the action
upon the brush-holder was not in practice sufficiently rapid when the
load changed suddenly or largely, and sparking therefore occurred
during the interval of lag.

Other engineers continued to work in order to improve compensating winding, and in 1903 Mr. Parsons and Mr. Stoney made
experiments which established that excessive overloading was possible
without sparking when the ratio of compensating ampere-turns to
armature ampere-turns was as high as two or two and a-half.
The compensating winding is chiefly concentrated on the polar
faces, and its extent suffices to neutralise the distorting effect of the
current in the armature windings on the field and, at the same
time, to provide a commutating field for reversing without sparking
the current in the coil undergoing commutation. Owing to the

[1] Patent No. 9203, A.D. 1900.

2 B

greater width of this commutating field, the brushes need not be so accurately set; in other words, there is considerable range of sparkless commutation at all loads.

Since this method of compensating involves no iron commutating pole it has the advantage that there is no self-induction to cause time lag at sudden changes of load; as the field in the gap between the poles is entirely in air it instantly responds to changes of current in the compensating winding, and thus the sparking often seen, with commutating poles, when there is a sudden change of load, is avoided

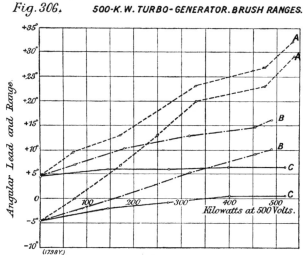

Fig. 306.　　500-K.W. TURBO-GENERATOR. BRUSH RANGES.

and the risk of a flash-over largely reduced. Moreover, since there is no iron to become saturated, the commutating field is always strictly proportional to the load, and thus sparking caused by saturation at certain critical loads is avoided. This advantage is felt still more when the dynamo is required to give varying voltages and to commutate well at any of them without any adjustment of the compensating winding by diverters or otherwise.

The effect of the system was clearly demonstrated in a paper, read by Mr. Gerald Stoney and Mr. A. H. Law, from which we have taken much information.[1] The authors give a diagram, reproduced on this page (Fig. 306), with commutating curves taken off

[1] "Journal of the Proceedings of the Institution of Electrical Engineers" (1908), Part 191, vol. xli.

exactly similar machines with varying amounts of compensation winding. Each pair of curves represents the limits of brush movement expressed in degrees within which the commutation was sparkless.

The two curves marked A A were taken on a 500-kilowatt machine, with the compensating winding out of action. It will be seen that only about 120 kilowatts could be carried without movement of the brushes, while at about 300 kilowatts the limits became very narrow, and beyond this no satisfactory commutating position could be found.

Curves B B show the limits with the ratio of compensation of 1.6, *i.e.*, ampere-turns on field to ampere-turns on armature. In this case 300 kilowatts could be carried without movement of the brushes, and, though the limits narrowed slightly, considerable overloads could be carried without sparking, if the brushes were slightly rocked forward. It may be of interest to state that the machine from which these results were taken was one of the first to be fitted with compensating winding, and as a slight brush movement was required up to full load it was also fitted with automatic brush gear.

The curves C C show an exactly similar machine with compensation of about 2.2 times the armature ampere-turns, and here it will be seen that there is no brush movement, and that, were the machine used as a motor, it would be quite reversible as far as the commutation is concerned, and that it can carry safely heavy overloads.

With this compensating winding there is a freedom from sparking with fixed brushes and fewer commutator segments can safely be used for any specified voltage, and more ampere-turns can be put on an armature of given diameter. The electrical output is greater for a stated speed, or the speed is higher for a given output. Before compensation was adopted, 500-kilowatt at 1800 revolutions per minute was the extreme limit in output for an armature at that speed, and not more than 10 per cent. overload could be carried. Now Messrs. Parsons have machines of 500-kilowatt running at 3000 revolutions per minute, and 750-kilowatt at

2000 revolutions per minute, carrying 25 or 30 per cent. overload without sparking, and 50 per cent. overload for a short time without injurious sparking. This has enabled the cost of the turbine to be largely reduced, and the steam consumption to be lowered by some 10 per cent. No difficulty has been experienced in running these compensated dynamos in parallel with one another or with other types, although in all cases they are self-exciting, no separate exciter being used.

In continuous-current dynamos a great improvement has been effected in recent years by the introduction of carbon brushes. The early slow-speed dynamos passed through a similar stage of evolution beginning with metal brushes, but gradually the design was altered and carbon brushes exclusively adopted.

For turbo-generators very special qualities of brush have to be used in order to give satisfactory working with the high peripheral speeds which are adopted, and it has also been necessary to design very special types of brush-holder. The brush-holder adopted by Messrs. Parsons consists of a box for holding the carbon and a plunger actuated by a spiral spring. This brush socket is extremely simple and adjustable in every way, and has enabled carbon brushes to be fitted to a large number of plants which were originally designed for working with a metal brush.

An interesting development in continuous-current armatures of new design has recently been introduced by Messrs. Parsons. Fig. 307, on Plate CXXIX., facing this page, is a perspective view of an armature of this type having an output of 1000 kilowatts at 1800 revolutions per minute at from 500 to 550 volts. Fig. 309 is a section of a portion of the core of this armature.

In this design the well-known advantages of the surface-wound continuous-current armature are retained, and at the same time the use of binding wire for holding the conductors in place is obviated. This is effected by locking a portion of each conductor into slots punched in the periphery of the iron core. The conductors are formed with a projecting portion which engages with the slots, the shape of the conductors being such that

Fig. 307. Continuous-Current Armature for 1000-Kilowatt Machine at from 500 to 550 Volts.

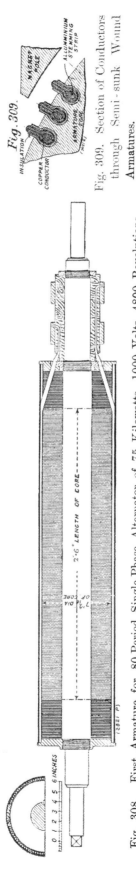

Fig. 309. Section of Conductors through Semi-sunk Wound Armatures.

Fig. 308. First Armature for 80-Period Single-Phase Alternator of 75 Kilowatts, 1000 Volts, 4800 Revolutions.

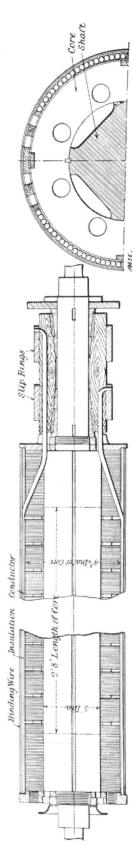

Fig. 310. 150-Kilowatt Single-Phase Turbo-Alternator, showing Air-Cooling Ducts.

Plate CXXIX.—Armatures.

they can be inserted in an axial direction, but are firmly held in a radial direction against the action of centrifugal force.

In the type of armature illustrated on Plate CXXIX., the method of holding the end windings is also novel. The usual bronze caps or binding wires are replaced by a number of bolts radially screwed into a central steel hub. This design has many advantages; it is of great mechanical strength, and enables the armature to be very thoroughly ventilated. It will be seen that every conductor, since it projects into the air gap, is very efficiently cooled, and, unlike other types of armature, there are no internal portions which can become overheated while the surface of the armature has apparently only a moderate temperature rise.

In designing this armature it was expected that exceptional results would be obtained as regards the commutation, owing to the fact that each conductor is only partially embedded in the iron core. A little consideration will show that the embedded portion of the conductor has a higher coefficient of self-induction than the portion projecting into the air-gap, and as a result the reversal of the current in the conductor at the moment of commutation is able to take place in a particularly easy manner. The portion of the conductor external to the core acts in a manner somewhat similar to the kicking coil of a field-magnet breaking switch, and entirely prevents any abnormal rise of voltage in the conductor at the moment of commutation. The expectations referred to have been very fully realised in actual practice, and it has been found that armatures of this type can be overloaded to a very remarkable degree as compared with the old surface-wound armature, while a comparison with the completely embedded type of winding is even more in their favour. On Plate CXXIX. there is a section through some of the conductors on the core of an armature of this type. The central portion of the conductor, which is of aluminium, is put in from the end in short lengths, and stemmed up tightly, causing the split conductor to open out and wedge itself very firmly in the slot, thus avoiding any possibility of movement.

Turbo - alternators are as acceptable to the electric station engineer as turbo-generators, and since the first was produced in

1889 many have been manufactured, ranging up to 6000-kilowatt capacity. The earliest machines, of 75-kilowatt capacity, for the Newcastle and District Electric Lighting Company, Limited, are illustrated on Plate CXXIX., facing page 188, and are of the revolving armature type, depending entirely on external ventilation for the dissipation of heat. They worked satisfactorily for fifteen years, and were only discarded then because larger machines were required.

In later and larger machines a square spindle was employed, with radial ducts and holes through the winding to pass air for cooling, as shown in Figs. 310 and 311 on Plate CXXIX., facing page 188, illustrating a 150-kilowatt machine running at 4800 revolutions per minute, and giving single-phase current at 80 periods and 2000 volts. By 1894, when the turbo-alternators had increased to 350 kilowatts capacity for 1000 volts, 100 periods, and 3000 revolutions per minute, twelve holes were made throughout the whole length of the armature, and air was induced through these by a fan; but no radial cooling was arranged for.

The earlier of these machines were surface-wound, but, owing particularly to the eddy currents developed in the conductors, which were large and insufficiently laminated, there was considerable trouble from heating, and they were subsequently altered to a tunnel winding with sixty holes, $\frac{3}{4}$ in. in diameter, forty of which were filled, each with one conductor, the remaining twenty holes being empty, as the alternator was single-phase. To this type also belong the Elberfeld machines, which were made in 1899, and already referred to in a preceding chapter (page 63 *et seq.*), as well as many others, ranging up to 1500 kilowatts capacity, and up to 4000 volts single, two, and three phase.

As the voltage was increased difficulties arose owing to the crossing of the windings at the end of a revolving polyphase armature, although the iron losses, especially in two-pole plants, were only about one-third those with a revolving field. The real trouble came when it was attempted to make a 1500-kilowatt 40-periodicity four-pole three-phase alternator with a revolving armature running at 1200 revolutions per minute, and a voltage

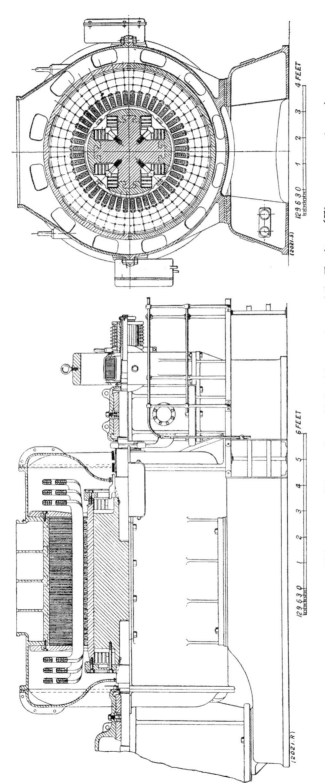

Plate CXXX.—4000=Kilowatt 3=Phase 6000=Volt Alternator with Exciter (Figs. 312 and 313).

Fig. 314. Armature of 6000-Kilowatt Machine.

Fig. 315. 4000-Kilowatt Stator ready for Dispatch on Truck to Carville Station of Newcastle-on-Tyne Electric Supply Company, Limited.

Plate CXXXI.—Latest Type of Armature and Stator.

of 6000.   It was found impossible to insulate sufficiently the end
windings, and at the same time to get rid of the heat generated,
and as a result the modern type of revolving field was adopted.

In revolving-field high-speed alternators of various constructions
there is comparatively little difference in the design of the stationary
or high-tension element.   This follows low-speed practice fairly closely,
though the external appearance is widely different owing to the few
poles, small diameters, and greater length of the high-speed plant.
It has, however, been found necessary to introduce a special system
of brackets or clamping rings to hold the end windings against the
enormous forces developed in case of a short circuit or other serious
disturbance on a system.   In the case of several alternators running
in parallel, these forces may be so great as to necessitate every
precaution in order to prevent movement of the coils, and no make
of turbo-alternator has been free from trouble from this cause,
though in recent designs the coils are so stayed as to be proof
against the most severe short circuits, and the necessity of thus
supporting the end windings as firmly as possible is now recognised
by all makers of high-speed alternators.

Figs. 312 and 313, on Plate CXXX., facing page 190, show
a longitudinal and cross-section of the 4000-kilowatt three-phase
6000-volt alternators installed at Carville Power Station, Newcastle-
on-Tyne.   The stators are split along the horizontal diameter,
and so arranged that the top half can be lifted by breaking only
four electrical joints.   This construction has the advantage for plants
of large size that the rotor can be lifted out direct, but it rather
complicates the end windings.   It has been found also to increase
the iron losses, and so all the most recent stators are not split,
but built in one piece.

Messrs. Parsons have constructed two, four, and six-pole rotors
of the salient-pole type.   Among them may be mentioned two four-
pole three-phase alternators of 6000 kilowatts, running at 750 revolutions per minute, which were supplied in 1909 to the New South
Wales Government Tramways, Sydney.   The stators are wound for
6600 volts, and are made in halves.

A number of alternators have recently been built with rotors of the cylindrical type, which offer some mechanical advantages. The largest of these are two of 7500-kilowatt capacity, constructed for the Randfontein Estates Gold Mining Company, of South Africa. These have six poles, and run at 1000 revolutions per minute, the stators being made in one piece. Fig. 314, on Plate CXXXI., facing page 191, shows a completed rotor of the latest type, and Fig. 315, on the same Plate, a completed stator. Fig. 316, on Plate CXXXII., facing this page, shows one of the latest 6000-kilowatt alternators with exciter.

As in the case of continuous-current machines, the object of the designer is to get the maximum output at the highest possible speed. The strengths of the materials limit the diameter and length of the rotor, and therefore both the available flux and ampere-turns. This, again, limits in the stator the output for a certain inherent regulation. If, however, any method of compounding the alternator to improve the regulation is adopted, much larger outputs can be obtained from a given weight of alternator. Many methods of doing this have been proposed, but most of them depend on either commutating a portion of the alternating current or causing certain parts to become saturated as the load increases, or upon varying the effective resistance of the winding on the exciter magnets. The first is objectionable on account of the trouble given by commutators, the second because the compounding varies with varying voltage, and the third is generally too slow to allow for rapid changes of load. A method of compounding has recently been brought out[1] which avoids these difficulties. The exciter is provided with an alternative path for the magnetism in parallel with its armature called a "leakage path." This "leakage" path is provided with a winding, which is in series with the main stator winding, and as the alternating current in this rises it chokes back the flow of magnetism, or, in other words, increases the magnetic resistance of the leakage path. Thus more flux is forced through the armature and, therefore, the voltage of the exciter rises and compensates for the tendency of the main voltage to drop. In a three-phase alternator there is generally one leakage path in each phase, and if the primary current is at high voltage a current

[1] Parsons' and Law's Patents No. 24141, A.D. 1905, and No. 13712, A.D. 1907.

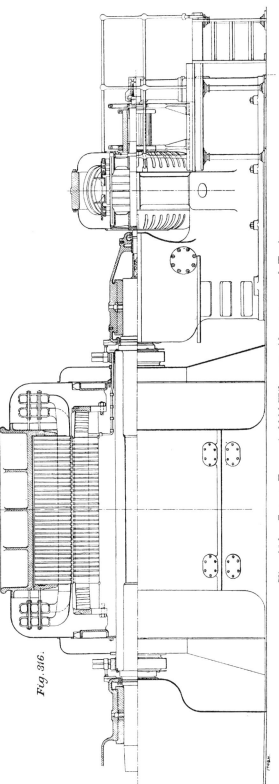

Fig. 316.

Fig. 316. Latest Type of 6000-Kilowatt Alternator and Exciter.

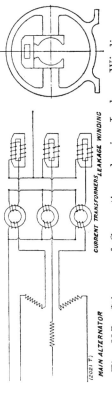

CURRENT TRANSFORMERS LEAKAGE WINDING

(2021 T)

MAIN ALTERNATOR

Fig. 317. Diagram of Connections showing Leakage Windings.

Plate CXXXII.—Latest Type of Alternator and Connections.

Fig. 319. Set for Underground Electric Railway Company of London, Lots Road Power Station.

Plate CXXXIII.—6000-Kilowatt Steam Turbine.

transformer is generally used. These leakage paths are easily adjusted for any desired amount of compounding ; or for any power factor. Fig. 317, on Plate CXXXII., shows diagrammatically this arrangement fitted to the exciter of a turbo-alternator, and Fig. 318, on this page, gives the comparative regulation on varying load with and without the leakage path in action.

This review of the development in design of electric generators

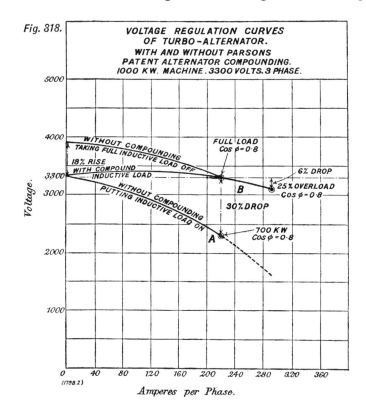

Fig. 318.

VOLTAGE REGULATION CURVES
OF TURBO-ALTERNATOR.
WITH AND WITHOUT PARSONS
PATENT ALTERNATOR COMPOUNDING.
1000 KW. MACHINE. 3300 VOLTS. 3 PHASE.

Amperes per Phase.

and alternators may fittingly be supplemented by a description of a typical large power turbine for driving such electric machines. As an example of modern practice in the design of the Parsons turbine the eight turbo-generators supplied to the Lots Road Power Station at Chelsea may be taken. Each of these turbines is rated at 6000 kilowatts, but, by means of the by-pass, the overload capacity is very large, and they are often called upon to develop 8000 to 9000 kilowatts for considerable periods of time. Indeed, momentarily

2 c

10,000 kilowatts has been reached, and at these great overloads there is only a very slight reduction in efficiency. When "exhausting to atmosphere," the turbines can take about eleven-twelfths of their nominal rating.

The turbines are divided into distinct high-pressure and low-pressure sections, with intermediate bearings between. There has thus been a reversion to the Elberfeld design, which is probably the most economical arrangement for turbines of very large power. The clearances at the high-pressure end, where the steam is dense, and the consequent leakage loss relatively large, can thus be kept smaller throughout than would be possible were the high-pressure and low-pressure sections assembled within a single casing. Further, the design of the casing is simplified, the size of the individual castings is kept down, and, high pressures and temperatures being confined to a casing of relatively moderate proportions, the risk of distortion in the latter is minimised. In the case of the turbines for Lots Road (of which illustrations are given on Plates CXXXIII. and CXXXIV., Figs. 319 and 320, facing pages 193 and 194), the necessity of fitting the machines to existing foundations somewhat hampered the design. Nevertheless, coupled to the existing generators, which have an efficiency of 95 per cent., the consumption on test proved to be but 14 lb. per kilowatt-hour, with steam supplied at 185 lb. boiler pressure and 140 deg. superheat, the vacuum being 28 in. with the barometer at 30 in.

A sectional view of the turbine is given on Plate CXXXIV., facing this page. As will be seen, the high-pressure rotor is a solid steel forging, and is thus absolutely free from any risk of expansion troubles and consequent slackening and shifting of parts. As the maximum diameter of this rotor is only about $3\frac{1}{2}$ ft., it was easily machined, transported, and erected. The low-pressure turbine was built much on marine lines. It has a drum 5 ft. 6 in. in diameter by $2\frac{1}{2}$ in. thick, and at 1000 revolutions per minute has a surface speed of 286 ft. per second, so that the centrifugal stresses, as in all modern turbines of the Parsons type, are moderate.

The two component parts of the turbine being independent, each

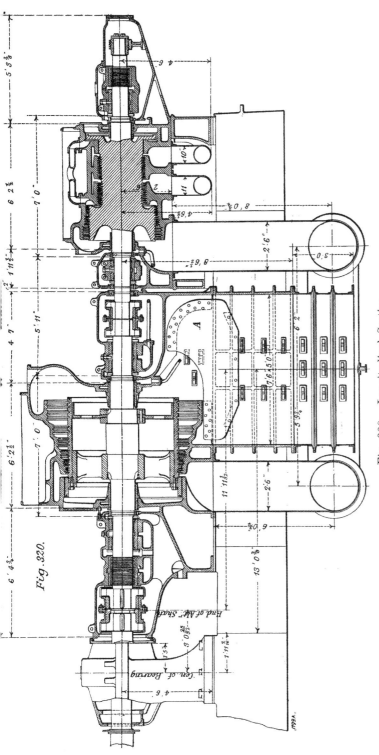

Fig. 320.

Fig. 320. Longitudinal Section.

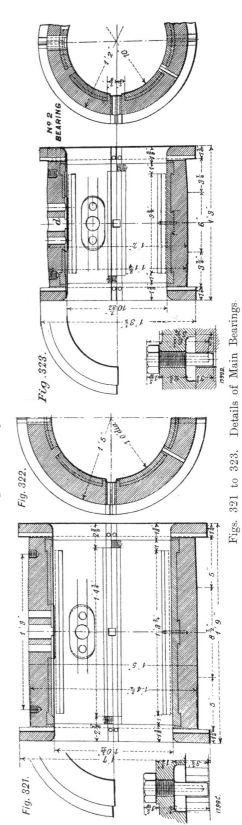

Fig. 323.

Fig. 322.

Fig. 321.

Figs. 321 to 323. Details of Main Bearings.

Plate CXXXIV.—Recently Completed 6000=Kilowatt Turbo=Generators.

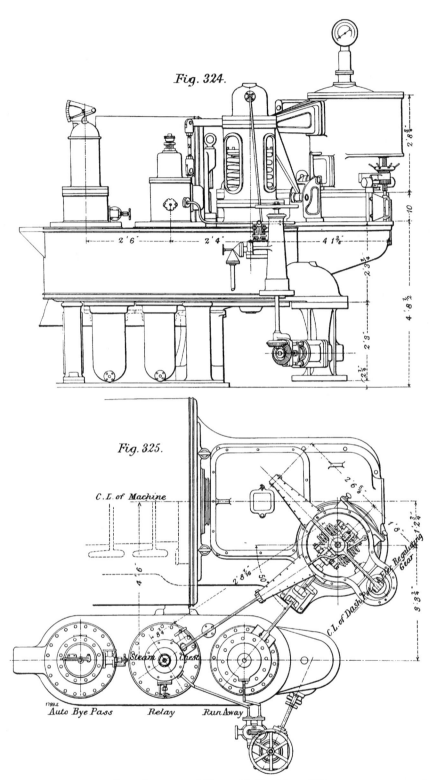

Fig. 324.

Fig. 325.

C.L. of Machine

Steam Chest

C.L. of Dash pot & Regulating Gear

Auto Bye Pass   Relay   Run Away

Plate CXXXV.—Valves for 6000-Kilowatt Turbine (Figs. 324 and 325).

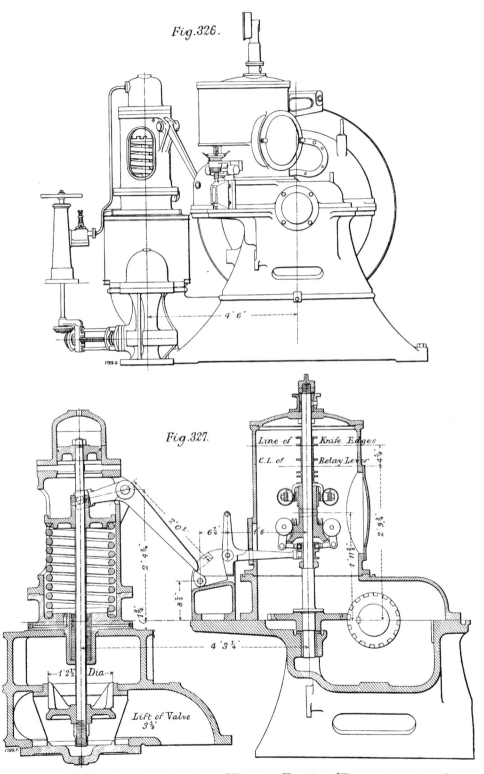

Fig. 326.

Fig. 327.

Line of ⟋ Knife Edges

C.L. of ⟋ Relay Lever

Lift of Valve
3⅝″

Plate CXXXVI.—Valves for 6000=Kilowatt Turbine (Figs. 326 and 327).

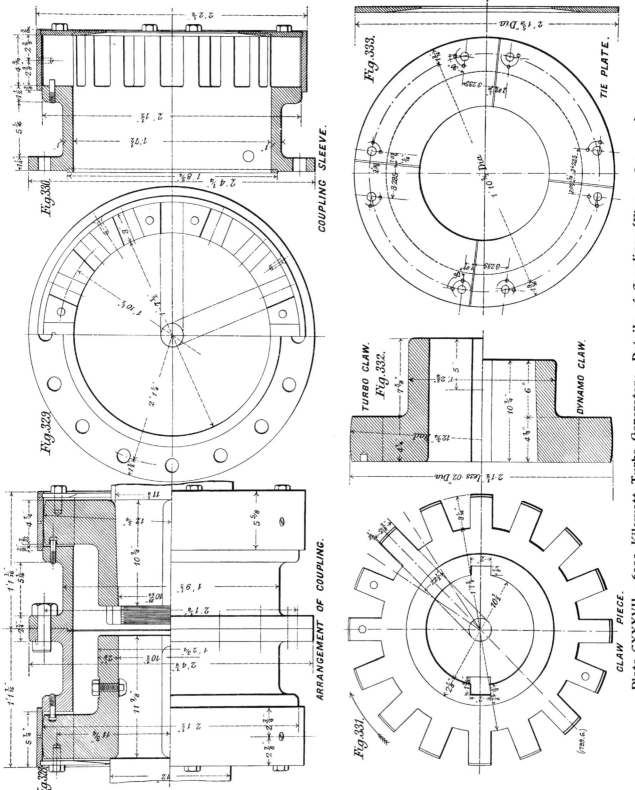

COUPLING SLEEVE.

Fig. 330.

Fig. 329.

Fig. 333.

TIE PLATE.

TURBO CLAW.

Fig. 332.

DYNAMO CLAW.

ARRANGEMENT OF COUPLING.

Fig. 328.

Fig. 331.

CLAW PIECE.

(739.G.)

Plate CXXXVII.—6000=Kilowatt Turbo=Generator; Details of Couplings (Figs. 328 to 333).

has its own thrust and adjusting block, and the dummy clearances
at the high-pressure end can therefore be adjusted quite independently
of those for the low-pressure drum, which, on account of its larger
diameter, requires a little more play. Owing to the necessity of
fitting existing foundations, the exhaust pipe from the turbine is
arranged centrally, as shown on Plate CXXXIV., facing page 194.
A transfer pipe, 2 ft. 6 in. in diameter, connects the exhaust from
the high-pressure turbine to the steam chest of the low-pressure
section. At full load the high-pressure section develops about four-
tenths of the total work.

Some details of the main bearings are reproduced in Figs. 322
and 323, on Plate CXXXIV., facing page 194. The shells are
of cast iron lined with white metal. Externally these shells are,
it will be seen, tapered off on each side of the centre, so that
they bear on their pedestals for a length of 8 in. only. This
narrow bearing serves the same purpose as the more expensive
spherical seatings frequently adopted, and gives more accurate
alignment. The oil supply is pumped into the bearing at the
middle of the line of the joint and at the top, and escapes at each
end of the bearing. In the top there are three holes, the two
outer being the main oil inlets, whilst the third takes the plug
shown, which serves, in the first place, to stop any possible rotation
of the brass in its seat, and, in the second place, as a thermometer
well, being drilled for this purpose as indicated.

The arrangement of the steam and governor valves is illustrated
in Plates CXXXV. and CXXXVI., between this and the preceding
pages. The main stop valve, which is of the Ferranti-Hopkinson
type, is placed just above the floor level, and is operated from a
horizontal hand-wheel by means of bevel gearing. The runaway
valve, the governor valve, and the automatic by-pass valve, are all
arranged within a single casting, which is located alongside the
high-pressure end of the turbine, as best shown in Fig. 325, on
Plate CXXXV.

The runaway valve, of which a section is given in Fig. 327, is
spring-loaded, and always tends to close, but in normal working is
prevented from doing so by a catch, worked by the runaway governor,

which engages with the longer end of a bell-crank lever as shown. Should the speed increase beyond the "emergency" limit, the catch is tripped, and the valve, closing automatically, shuts off the steam supply to the turbine. A dash pot arranged at the top of the valve spindle cushions the blow of the valve on its seat when closing.

Both governors are driven by worm gearing from the turbine spindle as usual, and are completely enclosed. They are spring-loaded, and the compression of the spring of the regulating governor can be adjusted so as to speed the turbine up or down, as circumstances may require.

This speed regulation of the turbine is effected direct from the station switchboard in the following manner :—By means of electro-magnets, one or other of the pawls carried on an arm kept in constant reciprocation by an eccentric, can be thrown in or out of gear with a ratchet wheel. The latter can in this way be given a step by step rotation in either direction. When moved one way it increases the load on the governor, and *vice versâ*. The main governor valve works on the "gust" and not the "throttle" principle. It is operated by a steam relay, the valve to which is kept in constant reciprocation about a "mean position," which alters as the governor sleeve moves up or down. This reciprocation of the relay valve is effected by a cam surface formed on the governor sleeve. According to the mean level of the relay valve, the steam is admitted in longer or shorter gusts to the turbine, thus controlling its speed. Since the governor valve is on this plan kept in constant motion " sticktion " is minimised, and prompt and certain adjustment of the steam supply to the load assured. The relay valve, which is that directly operated by the governor, is, of course, light, and offers but a small resistance to the action of the governor.

The by-pass valve, which admits steam direct to the third group of blades, is entirely automatic in action. It is a spring-loaded valve opening upwards, and normally kept closed by the pressure of steam at stop-valve pressure on the top of a piston. When, however, the steam-chest pressure which acts on the bottom of the piston rises above a certain pre-determined limit, the valve opens automatically,

Fig. 334. Fitting Up a 1000-Kilowatt Turbo-Generator Set in South Africa.

Fig. 335. 1000-Kilowatt Turbo-Generator Set Completed.

Plate CXXXVIII.—The Power Station of the Randfontein Estates Gold
Mining Company, Limited, Transvaal.

Plate CXXXIX.—2000=Kilowatt Direct=Current Turbo=Generator, 500–600 Volts, 1200 Revolutions (Fig. 336).

allowing steam to pass direct to the second group of blades. The point at which it opens is controlled by the spring. As already mentioned, the turbine is able to take extremely heavy overloads, and this is done quite automatically, and, as will be seen, by very simple means. In power-station work this automatic action of the by-pass is of extreme importance, since the changes of load occur with such rapidity that hand-operated by-pass valves are almost useless.

The flexible couplings by which the two sections of the turbine are coupled together, and the turbine as a whole to its generator, are illustrated in Figs. 328 to 333, Plate CXXXVII., facing page 195. They are of the claw type, which has long been used in all large plants by Messrs. Parsons and Co.

For the Randfontein Estates Gold Mining Company, Limited, Messrs. C. A. Parsons and Co. have built, at the Heaton Works, two complete turbo-alternators, of 6000-kilowatt capacity, and the first of these is now running. The order for units of this large size was the result of experience by the same mining company of several large turbo-alternators previously supplied to their extensive mining works in the Witwatersrand district of the Transvaal. These earlier machines generated three-phase current at 6600 volts and fifty periods, and were fitted with direct-coupled exciters arranged for a voltage of 100. Fig. 334, on Plate CXXXVIII., facing page 196, shows one of these 1000-kilowatt turbo-generators erected in the open during the dry season of 1907. The second machine was set to work in 1908, and is shown in position in the power-house in Fig. 335, on the same Plate. The condensers, and the air and circulating pumps, which are driven by a single motor operating at 550 volts, were manufactured by Messrs. Parsons, and the results have shown the total cost per unit at the switchboard to be only 0.384d., the price of coal being 10s. 5d. per ton.[1]

A typical direct-current turbo-generator of 2000 kilowatts is illustrated on Plate CXXXIX., facing this page. This engraving is illustrative of the Municipal Central Supply Station of the Borough

[1] See "ENGINEERING," vol. xc., page 735.

of St. Marylebone, in which there are five corresponding sets, in addition to four of 500-kilowatts, so that the total power in this station is 12,000 kilowatts. In this station the annual output is between ten and eleven million units.

One of the most successful electric power companies in this country is the Newcastle-upon-Tyne Electric Supply Company, Limited, for whose Carville station Messrs. C. A. Parsons and Co. supplied the first 3500-kilowatt turbo-alternators made at the Heaton Works, additional sets, of increased power, having been installed as necessity arose.

At this company's stations many interesting tests have from time to time been made, and we are indebted to Mr. J. T. Merz, Ph.D., D.C.L., chairman of the company, for the results of two typical comparative tests, which may here be given. The Neptune Bank Station, completed in 1900, is equipped with triple-expansion four-cylinder marine engines, of the most efficient type made at the time of its erection, as well as with Parsons turbines, direct-coupled in each case to three-phase alternators supplying energy at 5750 volts, 40 cycles. The diameters of the cylinders of the triple-expansion engines are :—21 in., 31 in., 34 in., and 34 in., with a stroke in each case of 3 ft. The full-load output is 800 kilowatts, at 100 revolutions per minute. The low-pressure cylinders are fitted with flat slide valves, while the high-pressure and the intermediate-pressure cylinders have trip valves. The cylinder clearances are approximately as follow :—

	Top.	Bottom.
High-pressure	10.6 per cent.	12.4 per cent.
Intermediate-pressure	11.0 ,,	12.5 ,,
Low-pressure...	8.4 ,,	8.4 ,,

in all cases in percentages of the volume swept through by the pistons. The turbines are of the Parsons type. The mean speed of the low-pressure blades is 270 ft. per second, at 1200 revolutions per minute, the mean diameter being 4 ft. $3\frac{5}{16}$ in. The normal load of the turbine is 1500 kilowatts, at 1200 revolutions per minute. Numerous tests have been carried out with a view to

Plate CXL.—View of Carville Station of the Newcastle-on-Tyne Electric Supply Company, Limited, where there are now Eight Sets of a Total of 32,000 Kilowatts (Fig. 337).

Plate CXLI.—375-Kilowatt Direct-Current Turbo-Generator, 110 Volts, as supplied for Ship Lighting (Fig. 338).

comparing the efficiency of the reciprocating and turbine plant, and the results of typical trials are appended :—

TABLE XII.—TESTS OF RECIPROCATING ENGINE AND TURBINES DRIVING THREE-PHASE ALTERNATORS AT NEPTUNE BANK STATION OF THE NEWCASTLE-UPON-TYNE ELECTRIC SUPPLY COMPANY, LIMITED.

Test No.	Average Load. Kilowatts	Average Steam Pressure in High Pressure Receiver. Pounds per square inch.	Average Steam Temperature. Deg. Fahr.	Average Vacuum. Inches.	Steam Consumption. Pounds per Kilowatt per Hour.	Approximate Thermal Efficiency. Per Cent.	Approximate Efficiency Compared with Rankine Cycle. Per Cent.
	Results of Reciprocating Engines: December, 1901.						
1	646	195	483	26.3	20.2	13.5	52.2
2	780	194	484	25.2	20.7	13.2	52.5
	Results of Parsons Turbines: April, 1902.						
1	1531	185	455	26.7	18.4	15	57.7
2	990	185	452	27.4	20.3	13.6	50.8
3	1691	192	462	26	17.8	15.5	61.25
4	566	194	460	28.1	23	12	42.3
	Results of Parsons Turbines: June, 1902.						
1	1442	196	462	27	18	15.25	56.4
2	1484	194	463	26.6	18.1	15.2	56.7
3	1153	196	471	27	18.6	14.7	54
4	1015	197	470	27	19.8	13.8	51
5	692.5	197	469	27.4	21.5	12.75	45.7
6	714	196	462	27.6	21.4	12.8	45.7
7	360	199	464	28	25.4	10.8	37.6
8	No load test.	200	455	28.2	Steam used 2949 lb. per Hour.	—	—

Table XII. gives comparative data as to the efficiency of the reciprocating and turbine sets. It will be seen that at full load the steam consumption of the triple-expansion engine was 20.7 lb. per kilowatt per hour, and of the turbine 18.4 lb. in the test of April, 1902, and 18.1 lb. in June, 1902. At less than full load the reciprocating engine compared more favourably.

The Commission appointed to consider the question whether reciprocating engines or Parsons turbines should be adopted for the propulsion of the Cunard express liners " Mauretania " and " Lusitania," referred to on pages 110 and 119, made additional tests at variable loads and speeds on the same engines at the Neptune Bank Station. The results are set out in Table XIII. In this

TABLE XIII.—CUNARD COMMISSION TRIALS OF RECIPROCATING AND TURBINE ENGINES AT NEPTUNE BANK STATION.

Reciprocating Engine.

Test No.	Speed R.P.M.	Average Load. Kilowatts	Steam Pressure. Pounds per Sq. Inch.	Steam Pressure in High Pressure Receiver. Pounds per Sq. Inch.	Temperature in High Pressure Receiver. Deg. Fahr.	Average Vacuum. Inches.	Steam Consumption per Kilowatt per Hour.
1	100.8	798	203	174	482	23	24.22
2	87.9	589	204	141	476	24¼	23.48
3	74.5	316	201	92	457	25.6	23.42
4	52.5	96.5	204	46	425	24½	30.93

Turbine Engine.

Test No.	Speed R.P.M.	Average Load. Kilowatts	Steam Pressure at Turbine Stop Valve. Pounds per Sq. Inch.	Temperature at Turbine Stop Valve. Deg. Fahr.	Average Vacuum. Inches.	Steam Consumption per Kilowatt per Hour.
1	1183	1478	203	453	25.9	18.67
2	1110	1092	204	448	26.7	19.65
3	903	600	203	440	26.8	24.01
4	649	196	212	436	27.7	35.01
5	1210	1822	203	480	25.5	17.80

case the steam consumption at full load was 24.22 lb. per kilowatt per hour in the case of the reciprocating engine. The consumption of the turbine was 18.67 lb., and at about three-quarter load the comparative figures were 23.48 lb. and 19.65 lb. As Atlantic liners run at or about full speed, these results were of more significance than those got with the engines working at lower proportions of the full load. The general conclusions reached were that the consumption of steam

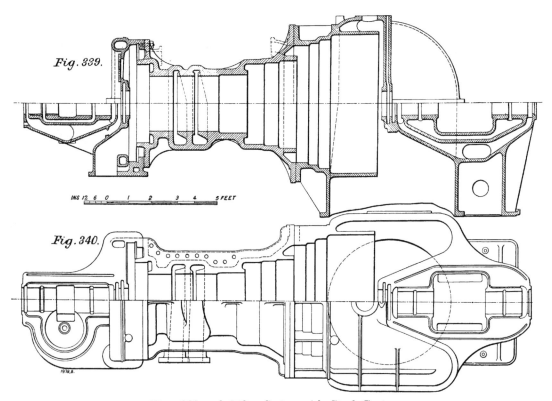

Fig. 339.

INS 12 6 0   1   2   3   4   5 FEET

Fig. 340.

Figs. 339 and 340.   Casing with Steel Centre.

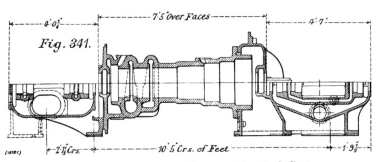

Fig. 341.

Fig. 341.   High-Pressure Casing with Steel Centre.

Plate CXLII.— Details of Turbines for High Degrees of Superheat.

per kilowatt per hour is lower for the turbine than for the reciprocating engine from full speed down to three-quarters of full speed,
the turbine being about 23 per cent. more economical at full speed,
16 per cent. more at nine-tenths of the full speed, and $2\frac{1}{2}$ per cent.
more at three-fourths of the full speed. At about half of the full
speed the turbine used about 13 per cent. more steam than the
reciprocating engine. Apart altogether from the influence of these
results upon the Cunard Commission, there is the fact that the
Newcastle-upon-Tyne Electric Supply Company have so largely
adopted the Parsons turbine in the extensions since made to their
generating plant at Neptune Bank and Carville.

The latest type fitted is nominally of 4000-kilowatt capacity,
generating three-phase current of 5750 volts of 40 periods. This
set, at normal load, has a water consumption for all purposes of
$13\frac{3}{4}$ lb. per kilowatt-hour, using steam superheated by 100 deg. Fahr.
When overloaded to give 5000 kilowatts, the water consumption for
all purposes is under 13.2 lb. per kilowatt-hour, while at 2000 kilowatts
the consumption is $14\frac{1}{2}$ lb. per kilowatt-hour. The output of this
station is between ten and eleven million units per annum, and the
coal and other fuel costs work out at only 0.20d., the oil, waste, water,
and stores at 0.04d. per unit, the wages at 0.08d., and the repairs and
maintenance at 0.08d., all per unit sold. These are exceedingly
favourable results, and are unexcelled by any station operated
by prime movers other than water turbines. The engraving on
Plate CXL., facing page 198, shows the Carville Station, but does
not include all of the Parsons turbo-generators fitted, the total
number now being eight, with a total capacity of 32,000 kilowatts.

Direct-current turbine-driven electric dynamos were adopted
for the lighting installations of the "Mauretania" and "Lusitania."
One of the eight sets made at the Heaton Works is illustrated on
Plate CXLI., facing page 199. As the view shows the top half of
the casing removed, the arrangement of expansion stages and system
of blading is well seen. These turbo-generators were designed each
to give 375 kilowatts at 110 volts when making 1200 revolutions
per minute. This slow speed was selected in order to make
absolutely certain that there should be no vibration or noise, and

the result has been entirely satisfactory. This advantage of the turbine is of great importance in passenger ships; piston engines or oil engines in many cases involve a most disturbing vibration and noise. The turbine sets in the Cunard liners were designed to exhaust to a condenser or to work against a back pressure of about 5 lb. per square inch. This latter condition of working enables the exhaust steam from the turbo-generators to be utilised in heating the boiler feed-water. When the ships are in port the turbines exhaust to the condensers.

And now we come to the later or rather the prospective developments. In land turbines in recent years very high degrees of superheat have often to be dealt with, the temperature reaching 600 deg. Fahr. and even more. Such temperatures as these have been found to exercise a detrimental effect on cast iron, not only altering the structure but also causing distortion of the metal. Such distortions in certain cases have given considerable trouble in land turbines, and, as a result, two alternative improvements have recently been adopted, and are illustrated on Plate CXLII., facing page 201.

In the first, as shown on Figs. 339 and 340, a steel casting is used for the centre part of the turbine, as long experience has proved that steel is not affected or altered in any way by the higher temperatures of superheated steam. The centre, being short and simple in form, involves no difficulties in casting it of steel; the low-pressure part of the turbine and the part at the high-pressure end not exposed to high temperatures are still made of cast iron. All the complicated portions of the casing are therefore confined to these parts. Thus under no conditions is cast iron in any of the more recent turbines subjected to higher temperatures than 350 or 400 deg. Fahr., which long experience has proved can have no detrimental effect whatsoever on cast iron.

This method of construction lends itself admirably to the tandem-compound turbine as adopted originally in the Elberfeld machine, and more recently in the 6000-kilowatt turbines for the Lots Road and other stations already described. All the parts, except the cast-iron ends of the small high-pressure turbine, can be made of steel. The

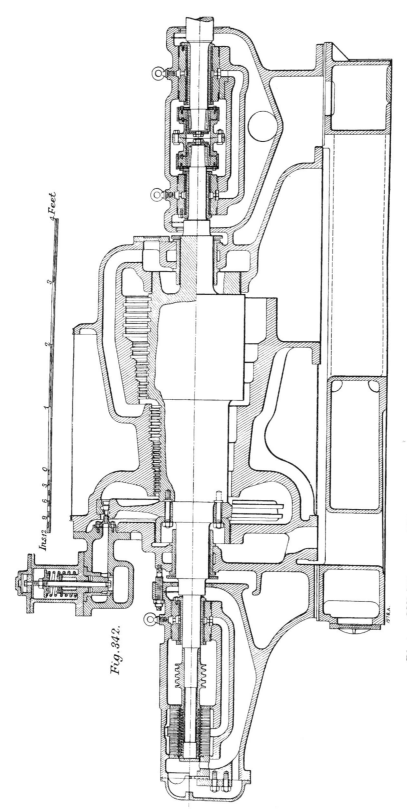

Fig. 342.

Ins.12 9 6 3 0 1 2 3 4 Feet

Plate CXLIII.—500=Kilowatt Parsons Impulse=Reaction Turbine (Fig. 342).

Plate CXLIV.—500=Kilowatt Parsons Impulse=Reaction Turbine (Fig. 343).

small high-pressure portion for a 3000-kilowatt tandem turbine at 2400 revolutions per minute is shown in Fig. 341.

The other alternative is to provide the turbine with an impulse wheel carrying, generally, two rows of blades, as illustrated in Plates CXLIII. and CXLIV. In a turbine of this type, with the wheel running at a surface speed of about 440 ft. per second, an initial steam temperature of 600 deg. Fahr. is so much reduced by the expansion in the nozzles that, even after allowing for the rise due to losses in the wheel, a final temperature is obtained no higher than 350 to 400 deg. Fahr.

This wheel, in combination with Parsons reaction blading, forms a simple, short, stiff spindle, giving, in moderate-sized turbines under usual conditions, about the same steam economy as the ordinary Parsons turbine. The relatively low efficiency of the impulse-bladed wheel in the one case finds its equivalent in the other in the large percentage of leakage in the short high-pressure reaction blades which are replaced by the wheel, so that the final result of the comparison is that both types have about the same efficiency. It is, however, doubtful whether in large sizes such an arrangement can compete for economy with the Parsons tandem turbine, where, in a small separate high-pressure casing, fine clearances with long blades can be obtained. The impulse-reaction system is largely favoured on the Continent, as is indicated in a succeeding chapter.

EXHAUST - STEAM TURBINES AND TURBO-AIR COMPRESSORS.

WHILE the propulsion of ships and the generation of electricity are the principal uses of steam turbines, there are other functions equally important and equally economically performed. In his patents of 1884 and 1894 Mr. Parsons clearly enunciated many different applications, and it is evidence of the accuracy of his anticipation of the full potentialities of his turbine that all of these alternative services are now being fulfilled by turbines, to the great benefit of industry. The more important of these (as noted on page 28 *ante*) refer to the compressing of air, to the driving of pumps, and to the working of turbines in series on the compound principle. It is fitting that the work done in achieving high efficiency in these miscellaneous duties should be here reviewed, before passing on to deal with the construction and management of turbines.

The application of the principle of the exhaust turbine calls for first consideration, as the exhaust steam from other prime movers is often utilised for driving rotary air-compressors, dynamos, pumps, &c., and thus there is converted to useful work heat energy otherwise and hitherto wasted. Typical examples are illustrated on the two succeeding plates, and described on page 208.

The exhaust-steam turbine, combined with the reciprocating engine, was first patented in 1894,[1] and the fundamental principles and the reason why economy is so certainly achieved have been elucidated in describing the condenser with a vacuum augmenter on page 58, and in reviewing the combination of reciprocating and turbine machinery in steamships on page 163 *et seq.* There is the further utility that the exhaust-steam turbine can take steam from many non-condensing piston engines, working intermittently, by the interposition of a steam reservoir or thermal storage accumulator between such engines and the turbine. Thus, at comparatively small capital cost and working charges, steam, which was formerly exhausted

[1] Patent No. 367, A.D. 1894.

Plate CXLV.—750-Kilowatt Exhaust-Steam Turbo-Generator at the Newburn Power Station of the Newcastle and District Electric Lighting Company, Limited, Operated by Exhaust Steam from Rolling Mills, &c., at Steel Works of Messrs. J. Spencer and Co., Limited (Fig. 344).

Plate CXLVI.—Exhaust=Steam Turbo=Blower Supplied to the Stafford Coal and Iron Company, Limited (Fig. 345).

into the atmosphere, is now utilised to drive turbo-generators giving up to 1500 kilowatts, and even more, to work turbo-blowers of 1000 horse-power, or to operate rolling mills through mechanical gearing—facts which establish the great gain conferred by the exhaust turbine in iron and steel works and other factories.

Still greater possibilities are in store for the system. In steel works the engines driving cogging and finishing rolling mills are so powerful that condensing plant is not often fitted, as the capital cost might be disproportionate to the gain because the engines are so irregular in their work. Thus the load in the case of large mills may vary from zero to 8000 horse-power many times within an hour, so that a condenser for the maximum output would seldom be working efficiently. On the other hand, when the steam is exhausted into a thermal accumulator for use in driving continuously an exhaust turbine working an electric generator, pump, or other machine, the average steam consumption of such a mill — about 50,000 lb. per hour—would suffice to keep in operation at full load an exhaust turbine and, for instance, an electrical generator with an output of 1500 kilowatts. This recovery of a valuable, and hitherto wasted, product does not involve the scrapping of existing engines and the introduction of new problems, as is the case with the substitution of the electrical method of driving mills. The first cost of the turbo-generator, too, is small. The same line of reasoning applies with equal force to the utilising of exhaust steam from blowing engines at iron works, winding engines at collieries, and many other prime movers working intermittently at other factories.

The turbine used corresponds exactly with those already described, and is designed to take steam at from 8 lb. to 16 lb. absolute per square inch and to use it expansively to 1 lb. absolute or even less. The exhaust turbine can be worked independently of the variations in the running of the other engines, because if the exhaust steam fails live steam is admitted to the thermal storage accumulator from the boiler through an ordinary reducing valve, set in motion automatically by the fall or rise in the steam pressure in the accumulator. Thus, when the reciprocating engines in the works become idle, with a consequent reduction in the volume of exhaust steam passing to the accumulator, the drop in pressure therein opens the valve to admit live steam from the boiler. On the resumption of work by the reciprocating engines and the resultant accumulation of exhaust steam in the reservoir, the valve admitting live steam closes, and the turbine is worked entirely by the exhaust steam. It follows, therefore, that the turbine only takes steam from the boilers when the requirements of the reciprocating engines slacken off or cease, so that no additional boiler power is required for exhaust turbines. Should it be desired to increase the electrical output from the exhaust turbines, live steam can be introduced by the adjustment of the same valve, in addition, and without reference, to the supply of exhaust steam. In many such cases, however, mixed-pressure turbines are used, as will be described later.

The exhaust-steam turbine is controlled by a centrifugal governor acting through a steam relay similar to that for ordinary turbo-generators. An additional safety centrifugal governor closes an emergency trip valve, which is incorporated with, but quite independent of, the governor valve, in the event of the revolutions exceeding by 10 per cent. the predetermined normal rate. The bearings are either of the tubular or solid pattern, according to the speed of the turbine, and are lubricated automatically by a small rotary pump driven off the turbine shaft. The condenser has the usual Parsons vacuum augmenter, the advantage of high vacua being, in the exhaust-steam turbine, about double that in the case of the ordinary high-pressure turbine, as has been explained on page 163 *ante*.

Where the supply of steam is fairly constant it is only necessary to fit a receiver to steady the pressure; an old Lancashire boiler shell serves the purpose admirably. But where there are great fluctuations or protracted stoppages in the working of reciprocating engines, and therefore in the supply of exhaust steam, two alternatives are available. A mixed-pressure turbine may be installed, and as it has high-pressure as well as exhaust-pressure blading, with an automatic governor for admitting boiler steam to the former, and exhaust steam (when available) to the latter, high economy is realised. To this we shall refer later. Meanwhile we may describe the other alternative, the fitting of a thermal regenerative chamber of some form to maintain the flow to a turbine for one or two minutes during the stoppage of the engine supplying the exhaust steam.

The function of the accumulator is to store heat during the working of the primary engines, and to give off heat during the period when such primary engines are stopped, so that there may be continuity of supply to, the turbines. In attaining this end there is utilised the principle that water, even although it may be at a lower temperature than 212 deg. Fahr., will still evaporate provided that there is maintained in the vessel containing it a pressure less than atmospheric pressure in proportion to the temperature drop. The steam exhausted into the accumulator-regenerator passes through a mass of water, violently circulated in order to cause the whole of the water in the vessel to participate in the interchange of heat. This circulation may be caused either by the incoming steam or by a separate pump, or otherwise. The water absorbs the heat units in the steam during the periods when the primary engines are working and exhausting into the accumulator; when the engines cease running the water commences to re-evaporate, owing to the reduction of pressure in the vessel.

The accumulator is usually in two compartments. It is surmounted by a steam receiver, which has for its object the separation of any oil and water that may be present in the steam, and the obviation of the violent shock which would otherwise be caused by the sudden admission of the steam to the water. The steam passes from the receiver to two ranges of branch pipes, and then enters special circulating tubes extending the whole length of the vessel. It has been proved by tests that a main engine stop of six minutes can be successfully dealt with, but stops of such length involve large cost in connection with the thermal reservoirs and provision for from one to two minutes is more usual. On the receiver there are two automatic relief valves, which prevent any excessive back pressure accumulating against the primary engines when the exhaust-steam turbines are working at less than full load. A pipe from the steam dome of the accumulator leads the steam to the stop valve of the low-pressure turbines.

On Plate CXLV., facing page 204, there is illustrated a typical installation of an exhaust-steam turbo-generator—a 750-kilowatt set at the Newburn Steel Works of Messrs. J. Spencer and Sons, Limited, Newburn-on-Tyne. The exhaust steam is derived from the rolling mill and other engines, and is utilised to produce continuous current at 500 volts, which is supplied to the Newcastle and District Electric Lighting Company, Limited, part of it being employed in Messrs. Spencers' Works for driving motors and for other work. The turbine runs at 1800 revolutions per minute, and takes exhaust steam at a little over atmospheric pressure. There is no thermal accumulator. The exhaust steam passes from the reciprocating engines through three reservoirs of Lancashire boiler form, and thence to the turbine. In the event of the supply of exhaust steam failing, live steam is supplied through a reducing valve on the turbine, as seen in the view on Plate CXLV. This turbine on test gave an efficiency of 71 per cent. of what is possible between the British thermal units and the electrical output—probably a record for any class of machine. It has been running at or over full load for two years without giving the slightest trouble.

Another typical installation—a turbo-blower designed to use exhaust steam at 18 lb. to 20 lb. absolute pressure—is illustrated on Plate CXLVI., facing page 205. It was built for the Stafford Coal and Iron Company. When tested at Messrs. Parsons' Heaton Works with a steam pressure of 3 lb. per square inch above atmosphere, and with a vacuum of 28 in. (bar. 30 in.), the total steam consumption was found to be 14,600 lb. per hour. The output was 14,000 cubic feet of free air at a pressure of 6 lb. per square inch. The principle of the blower will be described later.

Another proof of the great economy of the exhaust turbine is afforded by the results of an installation at Messrs. Guinness's Brewery in Dublin. This firm was supplied, in 1900, with a high-pressure turbine using saturated steam of 140 lb. per square inch for driving a 250-kilowatt generator. No condenser was fitted, as the exhaust steam—at about 11 lb. pressure above atmosphere—being free from oil or impurities, was led through low-pressure steam mains directly into the vats for heating purposes. Without

Plate CXLVII.—Exhaust=Steam Turbine at Messrs. Guinness's Brewery, Dublin (Fig. 346).

Plate CXLVIII.—"Mixed=Pressure" Turbine at Birtley Iron Works (Fig. 347).

a condenser this turbine has a water consumption at full load of 43.7 lb., and at three-quarter load of 49.4 lb. per kilowatt hour. It was considered that the steam, left over after heating the vats, could still be profitably employed in driving a low-pressure or exhaust steam turbine, and the machine illustrated on Plate CXLVII., facing page 208, was fitted. With a vacuum of only 26 in. the turbo-generator has an output of 250 kilowatts, the steam consumption at full load being 34.4 lb. and at three-quarter load 38.1 lb. per kilowatt hour.

It has been indicated that, with a thermal storage accumulator, the supply of exhaust steam may stop for from one to six minutes, according to the size of the accumulator, without live steam being drawn from the boilers, and that after this the live steam must be admitted to the accumulator. Where the stoppage in supply is too long to be met by the accumulator, what are known as "mixed-pressure" turbines are often fitted. In addition to the exhaust-pressure stages in these machines there is a high-pressure part, which revolves idly when exhaust steam is used; but an automatic arrangement admits live steam to this high-pressure part when the exhaust steam supply fails, and then the turbine, working on the compound principle, continues to drive the dynamo. This machine is most advantageous in cases where there are stoppages of mills, winding engines, and the like, during the day; and where electric supply may be required at nights or during the week end. It saves the waste of steam incurred in a simple exhaust turbine where a reducing valve is fitted for use during such stoppages.

This "mixed-pressure" turbine is a comparatively recent development, and the system is illustrated by the view on Plate CXLVIII., facing this page, which shows such a turbine, of 500-kilowatt capacity, driving a continuous-current dynamo of 500 volts, supplied to the Birtley Iron Works. The governor for automatically admitting live steam to the high-pressure part of the turbine, when exhaust steam is not available, is a special feature.

This governor[1] is sufficiently sensitive to come into operation, even if there is a temporary failure of the exhaust steam, for even

[1] Patent No. 6180, A.D. 1908.

2 E

a quarter of a minute, there being only about 1 per cent. variation
in speed between running on exhaust steam and on live steam.   In
special cases even this slight variation can be eliminated by the use
of a simple corrective device on the governing mechanism.   It
will be seen from Fig. 348 that there is a high-pressure relay and
a low-pressure relay, each of the usual type.   Both are worked from

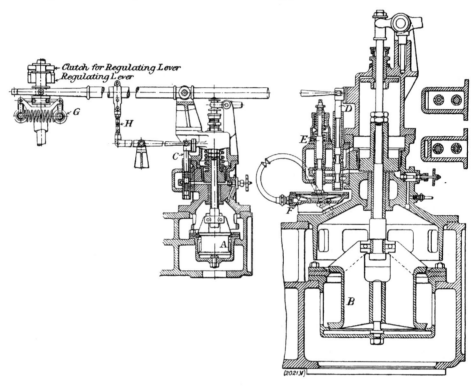

Fig. 348.   Governor of "Mixed-Pressure" Turbine for Automatically Opening or
Closing Steam Admission from a Receiver.

one governor, the relay valves being so set that steam is admitted
to the low-pressure stages before the high-pressure stages, and thus,
when the supply of exhaust steam fails, high-pressure steam is turned
on.   A is the high-pressure admission valve and B the low-pressure
valve.   Each is controlled by an ordinary relay gear worked by
plungers C and D which are connected to the governor G.   These
are so arranged that the valve A does not open until valve B is full
open and the speed decreasing, due to want of sufficient exhaust

steam. This is done by slightly altering the timing of the relay plungers C and D by the adjusting screw H. To prevent a vacuum being in the receiver when the supply of exhaust steam is not sufficient to drive the turbine, an arrangement is used in which a diaphragm F connected to the exhaust receiver works a relay plunger E, so that when the pressure below the diaphragm F—that is in the receiver—falls below a certain amount the plunger E drops and thus closes the double beat valve B and prevents the vacuum getting back from the turbine into the receiver. Air leaks are thus prevented. Otherwise, when the supply of exhaust steam failed, the pressure in the accumulator or the receiver would fall below atmospheric pressure, and air leaks would be caused, which would lower the vacuum, as such receivers or accumulators are never quite air-tight. Moreover, the receiver vessel might collapse were it subjected to full vacuum.

The turbine otherwise is of the ordinary type, but the low-pressure part is about double the capacity of an ordinary turbine working from boiler pressure to atmosphere, since the consumption in the case of exhaust steam is about double that of live steam. Although it might be expected that much loss would result from the high-pressure part moving idly when exhaust steam is available, the steam consumptions show that the loss is not greater than 3 to 4 per cent. A large number of these plants have been supplied, and the number is steadily increasing. There seems to be a great future for this type of turbine in places where there is an intermittent supply of exhaust steam. One great advantage is that it is not necessary in many cases to fit a thermal accumulator.

Turbines of the exhaust-steam or mixed-pressure types just described are largely used in all classes of factories, as well as the ordinary high and low-pressure types combined, not only for the generation of electricity but for other purposes. Amongst such duties that of air compression and furnace blowing deserve special notice, since here the applications involve special features in the design of the machine driven by the turbine with a great augmentation in efficiency.

Indeed such applications greatly influence, and are destined to

still further influence in the future, the economy of what may be
termed fundamental operations in engineering industry—the manufacture of iron.   It is only within the past ten or fifteen years that
the inexorable effects of competition have awakened the ironmaster
to the imperative need of the highest economy and of increasing
the yield from existing blast-furnace plant.   There is, for instance,
no longer a complacent satisfaction with the old steam-eating single-
cylinder beam blowing engine, even although the steam is generated
in boilers worked by waste gases.   There are other profit-yielding
uses for the gases, and steam economy becomes an element in the
situation.

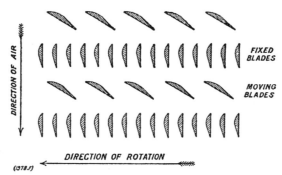

Fig. 350.   System of Blading in Turbo-Blowing Engine.

In 1901 Mr. Parsons adapted the principle of the turbine to
the design of a blowing engine,[1] which is extremely simple.   A typical
machine is illustrated by the section through the turbine and blower on
Plate CXLIX., facing this page.   Driven by a steam turbine of any
of the types described, is an air turbine, consisting of a casing within
which is a concentric drum mounted on bearings.   The blades or vanes
which propel the air are plano-convex in section, and are set in
rows at a suitable angle, as shown.   Between the rows of moving
blades are rows of guide blades inwardly projecting from the case.
These latter are also of the plano-convex section, and are set with
their plane surfaces parallel to the axis, as shown in Fig. 350,
on this page.   Their purpose is to assist the flow of air to the
moving blades and to stop its rotation after being acted on by
those blades.   The action of each row of moving blades adds a

[1] Patent No. 3060, A.D. 1901.

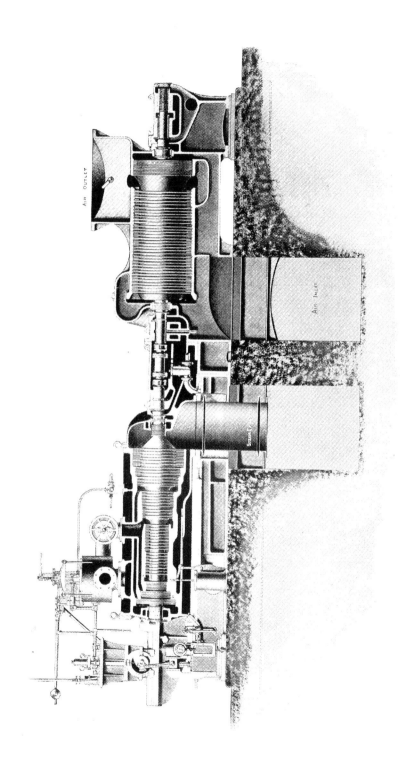

AIR OUTLET

AIR INLET

Plate CXLIX.—Section of Turbo=Blowing Engine (Fig. 349).

Plate CL.—Turbo-Blowing Engine at the Works of Messrs. Sir B. Samuelson and Co., Limited, Middlesbrough (Fig. 351).

little to the pressure, which increases gradually as the air flows from one end to the other of the turbine through the blades in the annular space between the drum and the casing. As in the steam turbine, balance pistons or dummies are provided for taking the end thrust of the air. The speed of rotation of a normal blast-furnace blowing engine is 3000 revolutions per minute, and the mean velocity of the air blades about 400 ft per second.

Great economy is attained owing to the high speed of rotation of the blower and the great range of expansion of the steam turbine with its vacuum-augmenting condenser. This blower soon commended itself, not only for its steam economy, but also for the facility with which the air supply could be controlled, and for the extreme steadiness of the blast pressure. The small area occupied, involving little space in crowded works and small buildings, the freedom from breakdown, the absence of wear and tear, and the small cost for maintenance, oil, and other "on cost" charges, were further reasons in favour of the system.

On Plate CL., facing this page, there is illustrated a typical blower, one in use at the works of Sir B. Samuelson and Co., Middlesbrough. This engine delivers 21,000 cubic feet of free air per minute against 10 lb. blast pressure, at the normal speed of 3000 revolutions per minute.

Another turbo-blowing engine for the copper-smelting furnaces of the Mount Lyell Mining Company's Works in Australia required only 5980 lb. of steam per hour when delivering 18,000 cubic feet of free air per minute against a pressure of 7 in. of mercury, which is equal to 255 adiabatic horse-power. Steam superheated to the extent of 70 deg. Fahr. was supplied at 140 lb. pressure per square inch, the vacuum at the cylinder exhaust being 27 in. mercury with the barometer at 30 in. The result gives a steam consumption per adiabatic horse-power hour of 23.5 lb., or an overall thermal efficiency of 34.3 per cent.

The view on Plate CLI., facing page 214, is specially interesting, as it shows the saving in space actually effected by the use of the turbo-blowing engine alongside of a reciprocating engine of the old type with blowing cylinders. The capacity of each set is the same—18,500 cubic feet of free air per minute against 10 lb. pressure. The comparison calls for little comment; the head-room required by the turbine blower is only about a fifth of that needed by the reciprocating engine. Another point is that the overhead crane need only be of about 5 tons capacity, since the parts of the turbine are so light. This installation at the works of Messrs. Wilsons, Pease, and Co., Middlesbrough, established a much higher overall thermal efficiency—37.4 per cent., although saturated steam was used, the pressure being about 150 lb. per square inch; and a vacuum of 27 in. was maintained. The steam consumption was only 22.5 lb. per adiabatic horse-power per hour.

There is possible, too, a great range in the volume and pressure of the air delivered. As in the case of the machine at Messrs. Samuelson and Co.'s works already described, a by-pass valve is fitted to admit steam at high pressure at an intermediate stage in the steam turbine, and it is thus possible considerably to increase the revolutions of the machine. In the case named the increase was from 3000 to 3600 revolutions per minute, with a corresponding augmentation in the air output or pressure, from 10 lb. to 15 lb. or even 19 lb. per square inch. The power required to drive the compressor varies approximately as the cube of the speed, so that in order to get 50 per cent. extra power it is not at all necessary to increase the speed of revolution by 50 per cent., but only in proportion to the cube root of the power. In this way it is possible to get 50 per cent. more power with only 500 extra revolutions per minute on the engine. The blast pressure may thus be increased very considerably, so that it can be readily understood that there is less liability of the furnace to develop "hangings" or "scaffolds." In any event, the furnace can easily be blown free should it choke.

The blower being rotary, the pressure is maintained continuously steady, whereas with the reciprocating-engine blower the blast pulsates with each stroke. There is thus a great advantage. Moreover, the steady running greatly simplifies and cheapens engine foundations. There are no valves to become heated or broken, and no need for cooling devices. The supervision is almost

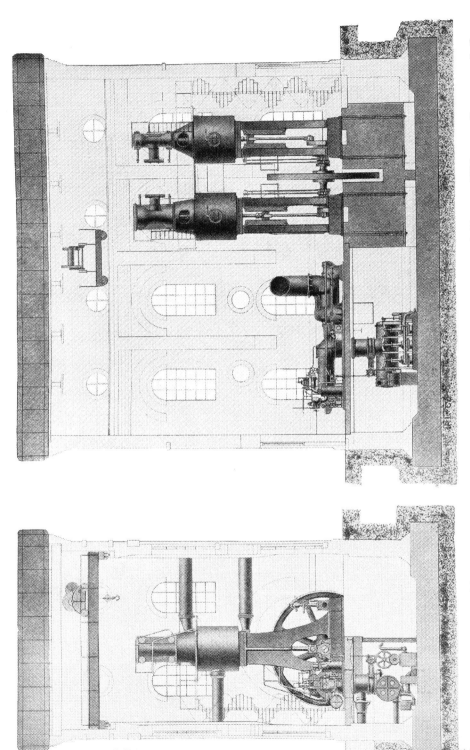

Plate CLI.—Installation of Turbo=Blowing Engine and Condenser at the Works of Messrs. Wilsons, Pease, and Co., Middlesbrough, showing the Saving of Space resulting from the Use of Turbo=Blowers instead of Reciprocating=Engine Blowers of Equal Output (Fig. 352.)

Plate CLII.—Turbo=Exhauster: View showing Internal Construction (Fig. 353).

negligible; one man can attend to three or four blowers. The lubrication is automatic, and in the case ot a machine to give 20,000 ft. of free air per minute against 10 lb. normal blast pressure, the consumption of oil is only about half a gallon per week, costing about 1s. 9d.—a small fraction of that required by reciprocating engines. The weight of such a plant is about 25 tons, or about one-fifth that of piston blowers. Finally, the first cost is not more than one half that involved by an ordinary blowing-engine ·installation when the buildings and foundations are taken into account.

These advantages, combined with the high economy, have resulted in a wide application of the system, and although it is only nine years since the first blower was introduced, there are now in use machines of a collective capacity of two and a-half million cubic feet of free air per minute.

Another application of equal efficiency is the exhauster for dealing with large volumes of air at low pressures.[1] The first machine of this type was introduced into the Coltness Iron Works in Scotland in 1904. The steam turbine is again the prime mover, while the exhauster, direct driven by it, consists of a shaft carrying three or more propellers, according to the output required. These are driven at speeds ranging up to 8000 revolutions per minute, and to pressures up to 100 in. water gauge. The view on Plate CLII., facing this page, shows the exhauster separately, and that on Plate CLIII., facing page 216, the exhauster directly coupled to the turbine.

It will be seen from both of these views that between the propellers there are arranged sets of stationary guide blades which are suitably curved to give maximum efficiency. These guides serve to take the spin out of the air or gas as it leaves the first propeller, and direct it on to the second one, where it receives a further compression, and so on to the third. When higher water gauges than 28 in. or 30 in. are required, even up to 100 in., it is only necessary to add to the number of propellers and guides. A noteworthy point in connection with these machines when dealing with blast-furnace gases is that, owing to the high velocity of the

[1] Patent No. 17802, A.D. 1904.

gas through the machine, there is no clogging with tar and other matter.

These exhausters have solved a difficulty which has been growing in intensity, especially in Scotland. The President of the West of Scotland Iron and Steel Institute in 1907 (Mr. T. B. Rogerson) narrated before the Institute the sequence of the consideration of the problem at his works—the Clyde Iron Works, Glasgow—so clearly that one cannot do better than quote his words [1] :—

"With the old engines they were continually having engineers in the shops effecting repairs, and they had to run these blowers at something like 100 revolutions per minute. The high speed at which they had to run the blower they found was knocking it to pieces, so they had to look around and do something. They considered first of all an ordinary steam engine with a piston, the same as worked on the reciprocating blowing engine for blowing air. They looked at this very carefully; but there were some things about it they did not like, because of the tar that was in the gas. The next thing they considered was exhausting the gases with a gas engine and an ordinary piston as arranged in the steam engine. This they considered very carefully, and had plans and specifications from at least three firms; but again they considered not the trouble with the gas cylinder, but the trouble with the exhauster cylinder—the trouble they expected they would get from tar in the gas. The blast-furnace gases of Scotland, in his opinion, were very different from those of Cleveland. Apparently, after they were washed they were very much cleaner; in fact, he had burned blast-furnace washed gas through an ordinary Welsbach burner, and had done it very well for a short time; but the burner had become choked up, and apparently the gas was perfectly free from dust. He believed it was, but it was not free from soot. They had, however, never been able to get to the bottom of it; but he believed that that had been the trouble. Where a blast-furnace gas engine had been blowing in Scotland, the trouble had not been with the engine, but with the gas, and they at the Clyde Iron Works considered this very carefully some years ago. He remembered going down to Middlesbrough some seven or eight years ago to see the blowing engine that was installed at Cochrane's This engine, he believed, was made in Seraing. When he saw it it was doing good work, and caused him to think that this was the thing to have. A gentleman whom he met at that time, and who had advocated gas engines for years, said to him : 'Be careful, Rogerson, what you are doing with gas engines and Scotch gas.' He had been very careful, and they were pleased to have got the blowing machines that they had; but he might say that he would have preferred to have had a gas engine for the sake of possible extensions at their works, where they could have used the gas in another department through a turbine; but he could not see his way clear with Scotch gas to introduce the gas engine."

The result was the fitting of two turbo-exhausters running at 8000 revolutions per minute, and each capable of dealing with

[1] "Journal of the West of Scotland Iron and Steel Institute," Session 1906-1907, vol. xiv., page 36.

Plate CLIII.—Turbo=Exhauster, showing Portions of the Exhauster Casing Removed so as to Expose to View the Propellers and Guide Blades (Fig. 354).

Plate CLIV.—Turbo-Exhausters in Engine Room at Clyde Iron Works, Tollcross (Fig. 355).

28,000 cubic feet of gas per minute against a total pressure difference of 28 in. water gauge. These are illustrated on Plate CLIV., facing this page. Mr. Rogerson, on the same occasion, said that these two turbo-exhausters, which they had had running then for something like twenty-two months, had given them no trouble whatever, and displaced five large exhausting machines of the blower description, which filled the engine house from one end to the other, and yet could not do the work efficiently in "pulling" the gas from, say, 160 tons of pig iron per day, whereas the two little turbo-exhausters were taking the gas from 260 tons of pig iron per day.

The same story might be told of many similar plants of equal capacities in many iron works, not only in this country but abroad. The measure of satisfaction is indicated by the number of repeat orders given for the plant. At the Coltness Iron Works four turbo-exhausters have replaced twelve exhausters of the revolving-drum type, the drums alone of which weighed together over 100 tons. These plants are arranged so that they can be used either for blowing or exhausting, and in several large iron works in Scotland have replaced the blowers of the revolving-drum type for dealing with the gases from the blast furnaces and pumping and forcing them through the ammonia-recovery plant. The capacity is about 30,000 cubic feet of gas per minute, and the suction or pressure about 1 lb. per square inch.

Turbo-blowers and compressors are now being made of the centrifugal type, the blower portion being very similar in design to the well-known high-lift centrifugal series pumps for water. On Plate CLV., facing page 218, there is shown in section a centrifugal blower capable of dealing with 4000 cubic feet free air per minute, and compressing to a final pressure of 20 lb. per square inch above atmosphere. The spaces between the various cells containing centrifugal fans are provided with water jackets, and it has been found from exhaustive tests that the efficiency obtained with this design of machine, when referred to adiabatic horse-power, can be made as high as 78 per cent., while the efficiency, referred to isothermal compression, can be got as high as 70 per cent.

On Plate CLVI., facing page 219, is an illustration taken from a photograph of this centrifugal blower, which has been constructed for the Burradon and Coxlodge Coal Company, Limited, Gosforth. This plant forms the preliminary compression stage for an existing reciprocating air compressor. The reciprocating compressor was originally used for compressing from atmospheric pressure up to 80 lb., and dealt with about 2000 cubic feet of free air per minute. By adding this turbo-compressor as a preliminary stage, the combined plant is now capable of compressing 4000 cubic feet of free air up to 80 lb., 20 lb. pressure being obtained from the turbine stage and the remainder of the pressure in the two air cylinders of the reciprocating engine. The power for driving the turbo-compressor is derived from the exhaust steam of the reciprocating portion of the installation, so that the output of the plant has been doubled without any increase in the coal consumption or number of boilers.

Centrifugal blowers have been supplied to the Electrolytic Smelting and Refining Company of Australia for dealing with 4000 cubic feet of free air per minute against pressures ranging from 8 lb. to 15 lb., and to the Sydney Gas Works for pumping town gas at the rate of 17,000 cubic feet per minute against 2 lb. pressure. A large centrifugal blast furnace blowing engine has been manufactured for Messrs. Sir B. Samuelson and Co., Limited, Newport Iron Works, Middlesbrough, for an output of 21,000 cubic feet of free air per minute against 8 lb. to 10 lb. pressure. It is driven by a non-condensing steam turbine, the exhaust steam from which will be used in a turbine driving an electric generator.

There are many similar applications of the turbine. In the year 1894 a turbine-driven forced-draught fan, consisting of a screw-propeller, was fitted on a yacht built by Messrs. Ramage and Ferguson, Limited, Leith, for supplying forced draught to the boilers. Many such installations have since been made, and turbines have been adopted similarly for driving large ventilating fans for collieries.

In 1894 also a turbine was coupled to a 6 in. Gwynne pump,[1]

[1] Patent No. 3024, A.D. 1895.

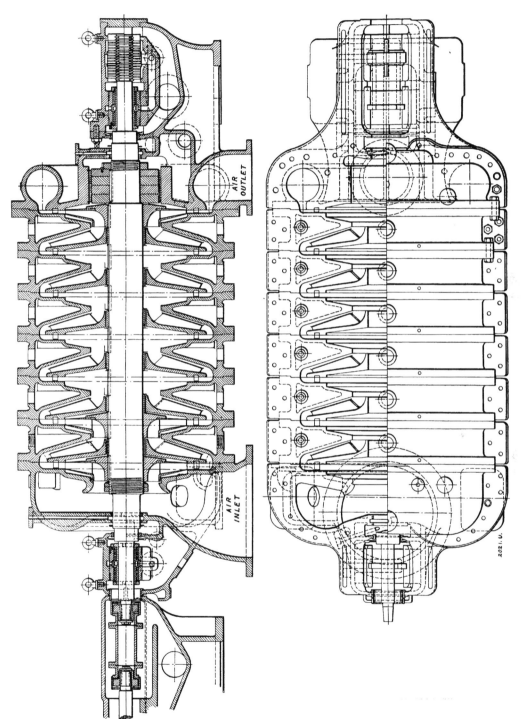

AIR OUTLET

AIR INLET

202 I. U.

Plate CLV.—Turbo=Air Compressor (Figs. 356 and 357).

Plate CLVI. - Centrifugal Blower for the Burradon and Coxlodge Company, Limited, Gosforth (Fig. 358)

and pressures up to 150 lb. per square inch were obtained with it at about 4500 revolutions per minute, and this was only limited by the casing of the pump failing to stand the pressure, as can be well understood, since the pump was only an ordinary one for 30 ft. to 40 ft. lift at 1200 revolutions per minute. This is probably the first case in which a high-speed, high-pressure pump was made. Screws were placed in front of the centrifugal impeller to feed the water in and to avoid cavitation, &c. Care had to be exercised in determining the ratio of pitch to diameter of these in order to prevent the formation of a vacuum. This also suggested the use of screws in series with increasing pitch to accelerate the flow of water.

Of later turbo-pumps two sets, supplied to the order of the New South Wales Government for the Sydney Water Works, are typical. The first set consists of a steam turbine with condenser coupled to three high-speed centrifugal pumps, which are arranged in separate casings in order that they may be coupled either in series or in parallel. When in series the first set is capable of lifting 1000 gallons of water per minute against 720 ft. head, while the second set is capable of lifting 2500 gallons of water per minute against 240 ft. head. In the case of the first set, the pumps are of the centrifugal type, and a view on Plate CLVII., facing page 220, is taken from a photograph of the finished machine. In the case of the second set, the pumps consist of a number of small propellers mounted together on a shaft with stationary guide blades between the propellers, and in principle they are very similar to the parallel-flow type of blower already described. Both of these sets of pumps give efficiencies of about 60 per cent.

The introduction of helical gearing, used between the turbine and the screw propeller shaft first in a launch and later in a cargo steamer, with most satisfactory results, as already described on page 174 *et seq.*, has greatly increased the possible applications of the turbine in industrial factories, as it renders it available for driving machinery which operates at a relatively low speed of rotation. Various such applications are proposed: indeed there seems little limit to the potentialities of the system, as it makes available the great steam economy of the turbine for low-speed drive with a negligible loss in transmission—about 2 per cent.

Perhaps the most suggestive illustration of such application is the use of a turbine for driving a rolling mill at Messrs. Dunlop's Calderbank Iron and Steel Works in Scotland. The turbine is of the mixed-pressure type, and capable of using either exhaust steam from the other rolling mill engines or high-pressure steam direct from the boilers. The speed is reduced in two steps. The first reduction gear is shown in Fig. 360, on Plate CLVII., and two views of the whole of the gears and of the turbine are reproduced on the two succeeding Plates.

The turbine itself runs at a speed of 2000 revolutions per minute, and in the first reduction gear this rate is reduced to 375 revolutions per minute. The second gear gives a further reduction down to 70 revolutions per minute, at which speed the mill rolls are driven. Between the second reduction gear and the rolling mill there is a shaft carrying a fly-wheel with a diameter of 23 ft., and weighing about 90 tons.

The mill, which is designed for rolling ships' plates, is of the three-high type, i.e., it has three rolls vertically arranged, and all three rotate continuously without reversing. The usual arrangement of tilting tables is provided and operated hydraulically for causing the plates to pass through between the top pair of rolls or the bottom pair of rolls. Although the turbine is only capable of giving an output of 750 brake horse-power, the actual power available at the mill shaft has been found to be as high as 4000 horse-power, the remainder of the power being supplied by the stored energy of the flywheel.

The governor of the turbine is of the usual type, and its effectiveness will be realised when it is stated that under ordinary working conditions full load is thrown on and off the turbine once in every 12 seconds. The use of exhaust steam enables the new mill to be run without the consumption of any live steam, and thus the power for this particular mill is got without adding to the coal bill or to the boiler plant in the works.

We have confined ourselves in this chapter to a consideration

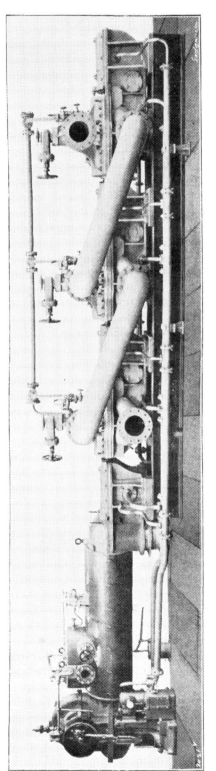

Fig. 359. Turbo-Pumping Plant for New South Wales.

Fig. 360. Rolling-Mill Gear, showing Helical Gear and Clutch.

Plate CLVII.—Two Recent Applications of Turbines.

Plate CLVIII.—Rolling=Mill Turbine, showing Helical Gear (Fig. 361).

Plate CLIX.—Rolling=Mill Turbine (Fig. 362).

of the applications of the turbine principle for various duties in British and Colonial factories. In the United States and in foreign countries the Parsons turbine is also extensively employed, and in the next chapter a few typical installations are described and illustrated.

PARSONS TURBINES FOR LAND PURPOSES
ON THE CONTINENT.

APART from some machines of a very early type, constructed in the nineties by the firm of Weyher, Richemond, and Co., of Paris, the Continental production of the Parsons steam turbine may be said to have begun with the formation of the Aktiengesellschaft für Dampfturbinen System Brown-Boveri-Parsons, and the granting of a manufacturing license to Messrs. Brown, Boveri, and Co., of Baden, Switzerland, and Mannheim, Germany, in 1900. This company undertook the exploitation of the system in Germany, France, Switzerland, Russia, Italy, and Belgium. In France the manufacture was placed in the hands of the Compagnie Electro-Mécanique, which installed works at Le Bourget, near Paris, and granted a sub-license to the Société de Fives-Lille ; in Belgium a license was granted to the Société Anonyme John Cockerill, of Seraing ; and in Italy arrangements were made by which the firm of Franco Tosi, of Legnano, acquired a manufacturing right.

A little later an Austrian company was created, named the Oesterreichische Dampfturbinen Gesellschaft, the construction being carried on in the works of the Erste Brünner Maschinenfabriks Gesellschaft, of Brünn. A manufacturing license has also been granted for Scandinavia to the firm of Burmeister and Wain, of Copenhagen.

The output of these Continental concerns soon attained a very large figure, as may be judged from the fact that up to the present date—a period covered by rather less than ten years—it amounts, for the countries exploited by the A.-G. für Dampfturbinen System Brown-Boveri-Parsons, and its sub-licensees, to about 1,800,000 horse-power, representing over a thousand machines. That this development has been a steadily increasing one may be judged from the fact that at the end of 1904 the total horse-power was only 340,000, whereas at the end of 1906 it was 875,000 horse-power, and at the end of 1908 it was 1,270,000 horse-power.

The Oesterreichische Dampfturbinen Gesellschaft was, up to the end of 1910, responsible for about 145 machines, aggregating 225,000 horse-power. This Company also has rapidly increased its rate of output; up to the end of 1907 only 95,000 horse-power had been delivered.

It would be tedious to describe in detail the numerous installations comprised in the above figures, but some of the more important may be briefly referred to.

The central station of Frankfort on the Main has four groups of 3000 kilowatts each and one of 3500 kilowatts. The first of these machines, which was started up in August, 1902, was the first large Continental-manufactured turbine. The machine works at 1360 revolutions per minute and monophase current at 3000 volts is generated.

Another very important station is that of Essen, where between 1905 and the present date seven machines have been ordered of 5000 and 7500 kilowatts each, the revolutions being 1000 per minute in each case, excepting that of the machine last ordered, which, while of the same power, runs at 1500 revolutions. All these generate tri-phase current at 5000 volts.

In the Oberspree Station, at Berlin, are three machines of 5000 kilowatts, 1000 revolutions, generating tri-phase current at 10,500 and 6000 volts.

The historical and successful Elberfeld machines, built by Messrs. C. A. Parsons and Co., which were the first Parsons turbines installed in Germany, have been followed by further orders for Continental - built machines of similar type. In addition to the Central Stations, the number of installations of Parsons turbines in Germany for manufacturing and mining purposes is also extremely large.

In France the most important station is that of the Société d'Electricité de Paris, at St. Denis. Indeed it is one of the most important stations in the world. A general view of the interior of the turbine engine room is given by Fig. 363, on Plate CLX., facing page 224, and a view of one of the 6000-kilowatt sets by Fig. 364, on the same Plate. There are ten turbo-alternators, each of 6000 kilowatts, 10,500 volts and 750 and 840 revolutions. The latest machine is a 11,000 kilowatt group for 750 revolutions.

The turbines in this station have been delivered between 1905 and the present date, and the success of this installation has no doubt contributed to the important order which has been received for the installation at St. Ouen for the Compagnie Parisienne de Distribution d'Electricité for five turbo-alternators, each of a normal power of 10,000 kilowatts, at 1250 revolutions, for generating bi-phase current at 12,300 volts. These machines are to take an overload up to 15,000 kilowatts. Amongst other prominent installations in France are those at Marseilles (containing four groups of 1800 kilowatts each), Puteaux, and Wasquehal.

In Belgium and Holland there are important stations at Amsterdam, Brussels, Liége, Ostend, and Rotterdam; in Denmark at Copenhagen and Frederiksberg, as well as some others.

In Italy the largest station is at Milan, where there are two stations, containing altogether twelve machines, varying in size from 1250 kilowatts to 5000 kilowatts. Other notable Italian stations having Parsons turbines are at Naples, Turin, Venice and Novare.

In Russia numerous machines have been installed at St. Petersburg and Moscow, and there are important stations at Kief and Baku, as well as a large number of machines supplied to mining, metallurgical, and manufacturing concerns.

Continental-built turbines have also been sent to Egypt, Portugal, France, Norway, Sweden, and the Argentine Republic. At Buenos Ayres, for instance, there is a station containing ten turbo-generators, ranging from 1500 to 11,000 kilowatts each.

The Oesterreichische Dampfturbinen Gesellschaft installs its turbines in Austro-Hungary and the Balkan States only. The most notable installation is that of the Vienna Municipal Electricity Works, where six groups of 6000 kilowatts each are already installed, and the seventh is in course of erection. There are two machines of the tandem-cylinder type, three of the single-cylinder type, and two of the combined impulse-and-reaction or disc-and-drum

Fig. 363. General View of Station containing Eleven Turbo-Alternators by the Cie Electro-Mecanique, Le Bourget (Seine).

Fig. 364. One of the 6000-Kilowatt Sets.

Plate CLX.—The St. Denis Power Station of the Société d'Electricité de Paris.

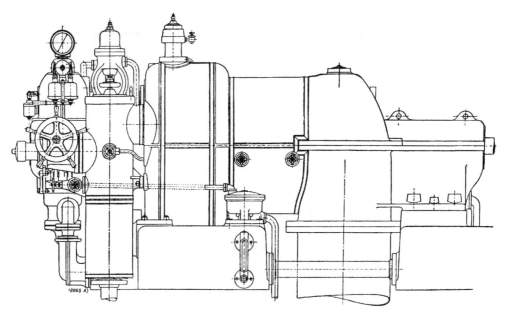

Fig. 366.   Elevation.

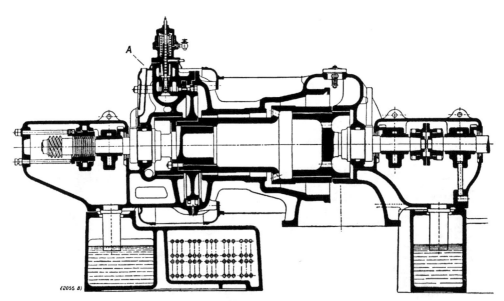

Fig. 367.   Longitudinal Section.

Plate CLXI.—Brown=Boveri=Parsons Impulse=Reaction Turbine.

type, to which reference was made in the chapter on electric generators and the turbines for driving them (see page 203).

The diagram, Fig. 365, shows the steam consumption, efficiency, and pressure curves for one of these machines—a 6000-kilowatt tandem-cylinder type, showing a very high thermal efficiency over a very wide range of powers.

As already stated, the impulse-reaction system is used to a considerable extent on the Continent, as will be seen when we

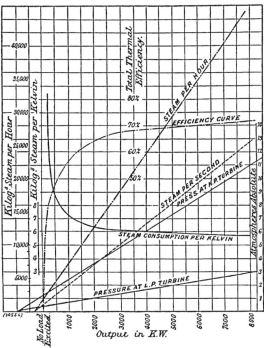

Fig. 365. Steam Consumption, Efficiency, and Pressure Curves of a 6000-Kilowatt Tandem-Cylinder Type Turbo-Generator in the Vienna Station.

state that up to the end of 1910 there was about 70,000 kilowatts already at work on the Continent.

Several illustrations are given representative of the most recent practice in respect of machines of this type. The illustrations on Plates CLXI. and CLXII., facing pages 225 and 226, clearly show the construction of the machine, as well as that of the governor and nozzle regulating gear. Fig. 371 shows a 1000-kilowatt turbine for 3000 revolutions (built by the firm of Brown, Boveri, and Co.), with the cover removed and the spindle exposed.

2 G

Fig. 372, on Plate CLXIII., facing page 227, shows a 2000-kilo-watt machine of this type for 1500 revolutions, built by the Oester-reichische Dampfturbinen Gesellschaft (Erste Brünner Maschinen-fabriks Gesellschaft), of Brünn, Moravia, for the Witkowitz Iron Works. Figs. 373 and 374, on the same Plate, show respectively a longitudinal section through this turbine and details of the governor and nozzle regulating arrangements.

The popularity of this type of machine on the Continent is largely to be ascribed to the practice prevailing there of using very highly superheated steam. In the disc-and-drum type of turbine the high temperature of the steam is confined to the part preceding the actual nozzle, and does not enter the body of the turbine or come in contact with moving parts. Thus a temperature of 600 deg. Fahr. in the steam pipe would be reduced to between 350 deg. Fahr. and 400 deg. Fahr. at the point of contact with the working parts of the turbine. The difficulties indicated can, however, be got over in the purely reaction turbine, as is proved on pages 202 and 231. In addition to the advantage attending the rapid drop of temperature, there is the gain of having the short, stiff spindle, which is so noticeable in the illustrations in Figs. 366, 367, 371 and 373. Moreover, the whole machine, being very short, is specially suitable where space is limited. It is not therefore surprising that this type of machine has been so well received.

It would be misleading to state that any considerable gain in efficiency is to be expected from the disc-and-drum turbine as compared with the reaction turbine. Under special conditions, as when machines are underspeeded for their power, including those intended for driving continuous-current generators, the disc-and-drum turbine does show an increased economy over purely reaction turbines. On the other hand, for large machines there is no gain in efficiency as compared with single-cylinder pure reaction turbines. For the very largest class of machine, where the very highest economy is sought by the use of two separate cylinders, irrespective of the somewhat higher first cost of this construction, the purely reaction turbine comes out with a higher efficiency than the disc-and-drum type.

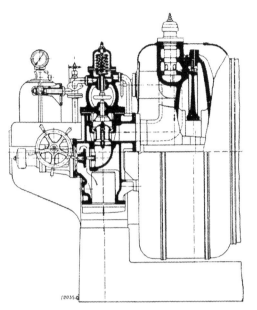

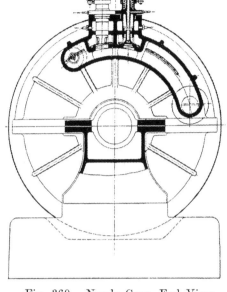

Fig. 368.   Governor and Nozzle-Regulating Gear.   Fig. 369.   Nozzle Gear, End View.

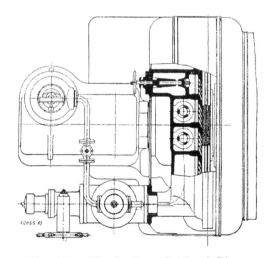

Fig. 370.   Nozzle Gear, Sectional Plan.   Fig. 371.   1000-Kilowatt Turbine with Cover Removed.

Plate CLXII.——Brown=Boveri=Parsons Impulse=Reaction Turbine.

Fig. 372. General View of 2000-Kilowatt Machine.

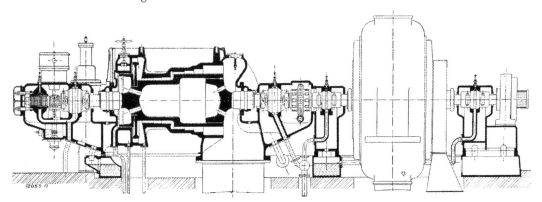

Fig. 373. Longitudinal Section through Turbine.

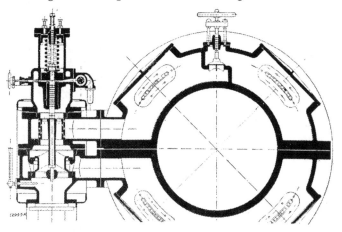

Fig. 374. Governor and Nozzle Regulating Gear.

Plate CLXIII.—The Austrian-Parsons Impulse-Reaction Turbine.

The matter may be put in a nutshell by pointing out that the impulse stage has an inherent thermal efficiency lower than that of reaction blading; but that, in some cases, it provides a method of working over a larger pressure range than it is always convenient to utilise in reaction blading. The comparison is therefore between blading of moderate thermal efficiency working over a large pressure range, and blading of high thermal efficiency working over a smaller pressure range.

Noteworthy installations on this system are at St. Denis, St. Ouen, Essen, Heidelberg, Paderborn, Rixdorf, Arnhem, Novare, Cairo, Bruay, Epinac, Louvroil, Beeringen, Heerlen, Moscow, Altona, Dresden, and many other places.

Turbines on this principle were constructed in England as far back as 1904 by the Westinghouse Company, but the design was abandoned by them in favour of the double-flow type with impulse blading in the centre of the length of the drum. It was taken up again later on the Continent by the Swiss and Austrian Works, where machines were completed in the year 1907.

THE MANUFACTURE OF TURBINES.

HAVING traced the evolution of the turbine through a long series of most interesting experiments, and reviewed the stages in its wide application on steamships for propulsion and electric lighting, and on land, at home and abroad, for driving electric generators, and for many other duties, it seems appropriate that a description should be given of the process of manufacture. This is particularly justified, as in some respects workshop methods had to undergo modification, and new manufacturing tools had to be introduced, with the advent of the turbine.

The very fine clearances necessary to efficiency demand an unusual degree of accuracy even in machines of large dimensions. This will be the more readily appreciated when it is noted that a rotor weighing as much as 120 tons has been finished with the precision necessary to work in a casing with a clearance of about one-tenth of an inch. Such a result requires not only precise turning, but extreme care in the design of the bearings, accurate proportioning of the rotor to obviate whipping, and casings of correct dimensions and thickness to prevent sagging; while in addition to all are the methods needful to overcome any possible inequality in contraction or expansion due to heat.

Thus, for instance, it was deemed prudent in the case of the earlier turbines to make practically solid patterns for the casings, in order to ensure thoroughly satisfactory results, alike as regards dimensions of casting and thickness of metal. As the ironfounders gained experience both in material and methods, the pattern-maker was able to reduce considerably the wood-work. Now patterns are often made of the "skeleton" type only, the boxes and boards, few in number, representing only the main contour and dimensions of each particular unit of the casing. Such patterns, however, leave more for the founder to do, and entail greater care and experience on his part to eliminate the possibility of foundry errors. Semi-solid patterns are therefore still made in many cases, retaining the main features of the solid finished pattern.

The iron castings for the casings, when received into the works from the foundry, are, of course, thoroughly sounded for surface defects, internally and externally. They are then put on the marking-off table and checked for dimensions and thickness, more especially for thickness, because some founders have a tendency to turn out castings considerably in excess of the designed thickness. This has to be counteracted, as the designed weight of the turbine is thereby increased. The parts of the turbine casing—upper and lower halves in the case of small turbines, or two upper and two lower sections, with a circumferential as well as a horizontal flanged joint, in large turbines—are next rough planed and drilled together, after which the casing is put on the boring machine and rough bored and grooved, $\frac{1}{16}$ in. being left on all diameters for final machining. The next operation is to test the casing under water pressure, at usually one and a-half times the initial working steam pressure of the turbine. After water testing the casing is annealed by steam, and in some cases by a special furnace, to release any strains in the casting.

The top and bottom halves are then fine planed on the joint, after which the casing is again put on the boring machine for final boring and grooving. Great care is necessary in setting both the cylinder and boring bar, so that when the final boring is completed the cylinder bores shall be perfectly true. Handy gauges are supplied from the standard gauge room to the men attending the boring mill and the large lathe for turning rotors, and all work is checked by special gaugers before it leaves the machines. Special laths, marked from the micrometer, are used in marking off the grooves for holding the blades, both in the casing and rotor, in order to ensure that the fixed and moving blades shall have correct end clearance. After fine boring and grooving, the casing is ready to be fitted with blades, to be described later. Fig. 375, on Plate CLXIV., facing the next page, shows the lower halves of two casings ready for blading.

During the construction of the casing the production and

assembling of the units of the rotor are in progress. In present-day practice these two—casing and rotor—seldom come together until the final operation of completing the turbine preparatory to its being tested under steam. The drum, or main tube of the rotor, is usually of forged steel, of about 30 to 35 tons quality. The drum can either be forged or pressed out of the solid ingot, or may be spun out of the ingot by special machines designed for that purpose. The wheels connecting the spindle ends of rotor shafts to the drums were originally made of cast steel, but in recent British Admiralty work, and in some merchant marine turbines, they are made of forged steel, being cut out of large solid ingots. This entails much work in their formation, but offers greater surety in the finished article. Cast-steel wheels are more generally used for mercantile work. The wheels are shrunk on to the rotor shafts, which are of forged steel of from 34 to 38 tons quality, and are generally bored hollow. The arms of the forward wheels—those at the steam admission end of the rotor —are usually made hollow, so that live steam may heat the wheel and enter a part of the rotor shaft through holes in the hub and shaft. This heating ensures a proper expansion of the wheel and rotor shaft, and is helpful in keeping a close and unvarying clearance at the dummies or steam baffles.

The procedure adopted in the building up of a marine rotor may be briefly described. The drums are usually received from the manufacturers rough machined. The first operation is to fine bore and finish them to length, and in the former process accurate turning is essential. The wheels, too, are fine bored, finished to length on the boss, and the outer ring rough turned. The wheels are then carefully balanced on knife edges, after which they are shrunk over the rotor ends. As a rule the rotor shaft is made $\frac{1}{500}$th to $\frac{1}{750}$th greater diameter than the bore of the wheels to facilitate the shrinking-on process. After the wheels have been shrunk over the rotor ends, the outer rims of the former are fine turned and the drum is shrunk over them. The drum is then drilled, and screwed rivets are fitted through the drum into the rim. The rotor is finally put into a large lathe to have the finishing, turning, and grooving done. The rotor is then ready for blading, as shown in Fig. 376 on Plate CLXIV.

Fig. 375. Casings Ready for Blading at Turbinia Works.

Fig. 376. The Finished Turning and Grooving of a Rotor at Turbinia Works.

Plate CLXIV.—The Manufacture of Casings and Rotors.

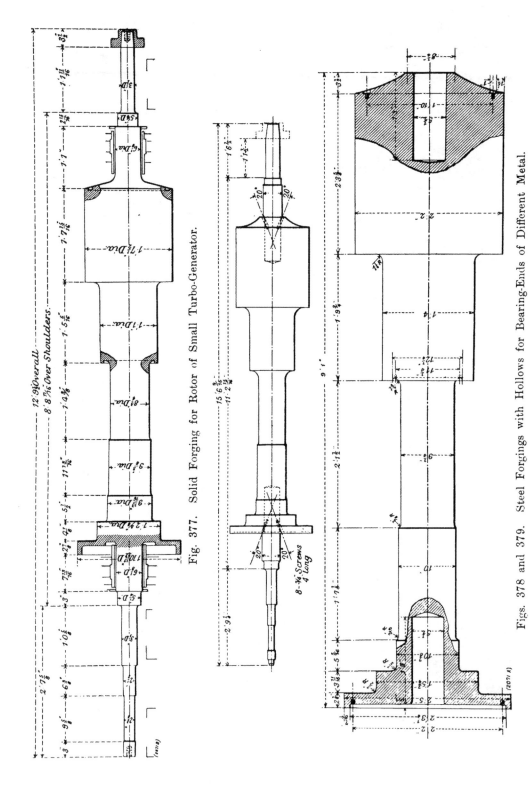

Fig. 377. Solid Forging for Rotor of Small Turbo-Generator.

Figs. 378 and 379. Steel Forgings with Hollows for Bearing-Ends of Different Metal.

Plate CLXV.—Forged Spindles for Rotors of Turbo-Generators.

In the case of rotors for high-speed turbines driving electric generators and blowers, the method of construction, as a rule, differs from that adopted in marine work, and varies according to the size of the rotor. For small rotors up to 500 or 750 kilowatts a solid forging is used, with the ends solid with the body, as shown in Fig. 377, on Plate CLXV., facing this page. In many cases, where it is desired to have the ends carrying the bearings of a different class of steel from the main body, these ends are made separate and forced into the solid body as in Figs. 378 and 379, on Plate CLXV. This formation is the most economical for small sizes, but becomes too heavy for large sizes, and then the main body is made of a steel casting and, in more recent times, of a steel forging, with the high-pressure end shrunk in and the low-pressure end supported in the body by a steel wheel, as shown in Figs. 380 to 384, on Plate CLXVI. The bodies of this type of rotor were formerly of steel castings, but improvements in forging have enabled them to be obtained at a reasonable price in forgings; these are infinitely preferable to steel castings, which are liable to serious defects such as blow-holes, flaws, &c.

In large sizes where a high degree of superheat is adopted there has been some trouble due to the high-pressure end coming loose, and to reduce this risk it is now made hollow and supplied with steam through the body, as shown in Figs. 380 and 381, on Plate CLXVI., facing page 232. Besides this, the end is strongly pinned into the body. Later it was considered preferable to have the end solid with the body, and after consultations with the leading manufacturers of large forgings in this country, the design as shown in Figs. 382 to 384, on Plate CLXVI., was adopted. In this case the high-pressure end is solid with the body, the low-pressure end is supported by a steel wheel, the low-pressure dummy consisting of a piece of boiler plate bolted on to the main body. It will thus be seen that no part subject to high temperature or high pressure has any joint in it, and thus all possibility of movement is prevented.

This design has been arrived at after many years of work and experiment, and it has only been made possible after many consultations with the large forge masters in this country.

In the case of combined impulse - reaction turbines, similar designs are adopted according to size, but in every case the large impulse wheel is made separate and bolted on to the body, as shown in the drawing of a 500-kilowatt impulse-reaction turbine reproduced on Plate CXLIII, facing page 202. The dummy is made of a separate forging. The boxes carrying the nozzles are made of cast steel, and are independent from the main cylinder casting. The boxes are divided into a number of sectors as may be required for the varying conditions of power desired. The sectors may be built of forged steel carrying the blades forming the nozzles, which blades are calked in a similar manner to the ordinary Parsons practice, or the sectors with nozzles may be cast complete.

And now we may deal with the interesting process of blading. The material used in the manufacture of the blades was dealt with on page 50. In the ordinary Parsons marine turbine the blades are now made of an alloy containing about seventy parts of copper and thirty parts of zinc. Special ingots of the material are cast and rolled to the required thickness, and are then cut to the required widths and lengths by splitting rolls. The strips formerly were passed through rolling mills to give them the approximate shape, and the finished curvature was produced on a draw-bench by passing the blade strip through dies having highly-polished surfaces. Recently, however, the rolling mills have been so perfected that the blades can be produced direct from the rolls without the intervention of the draw-bench. The experiments as to the most suitable form or curvature are reviewed on pages 29 and 46.

The strips for the distance or packing pieces put in the grooves between the blades are similarly formed. The lengths of these packing pieces correspond to the depths of the blade grooves. The cutting to length is done by means of small saws revolving at high speeds.

The strips of blading material are cut and formed by special stamping machines into the requisite lengths for the different expansions of any particular turbine. The strips of blade material pass through the rollers of a special press, where they are not only cut to length, but also have the root serrations and binding wire

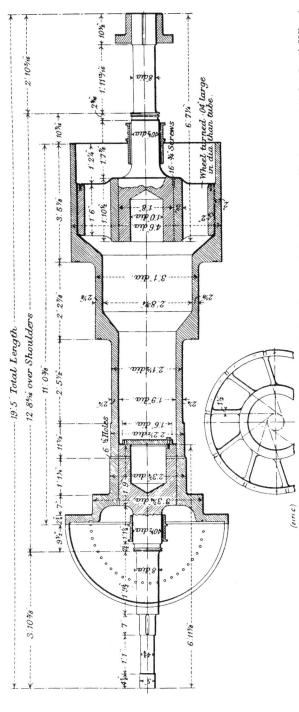

Figs. 380 and 381. Forged Hollow Steel Body, with High-Pressure End shrunk in and Low-Pressure End supported by Steel Wheel.

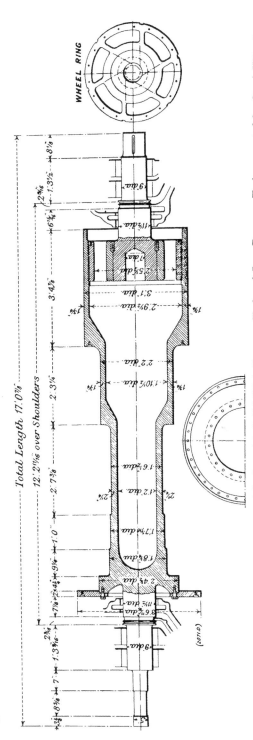

Figs. 382 to 384. Forged Hollow Steel Body, with Solid High-Pressure End and Low-Pressure End supported by Steel Wheel.

Plate CLXVI.—Forged Spindles for Rotors of Turbo-Generators.

Fig. 385. Cutting Blades into Lengths.

Fig. 386. Tipping the Edges of the Blades.

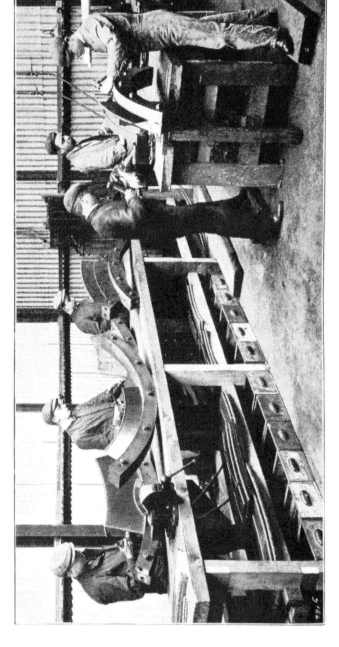

Fig. 387. Forming Segments of Casing and Rotor Blading.

Plate CLXVII.—Blading Turbines.

grooves on the edge formed at one operation. An illustration
of this is given in Fig. 385, on Plate CLXVII., facing this page.
The blades are then taken to small tipping machines, where the
tips are specially thinned. This thinning, as already explained,
has been of great service in cases where there has been accidental
contact of the blade with the proximate surface whilst running.
The possibility of stripping has thus been obviated. The blade is
inserted in a groove in the machine while a milling cutter is advanced
over the top. Fig. 386 illustrates the process of tipping the blades.

The blade pieces and packing pieces are generally packed and
stored in definite quantities until required.

Two systems of fitting or attaching the blades to the rotors
and casings are in vogue. The older system has come down from the
experiments described on page 48—now known as the "individual"
type, while the newer method is called the "segmental" or Rosary
system. The former consists of driving separately the "individual"
pieces, alternately blade and packing piece, round the grooves in
rotor and casing respectively until the ring is complete, and then,
by means of a suitable calking tool, of swelling out the packing
pieces between each blade. In the case of the "segmental" form
of blading the blade and packing pieces are assembled together in
sections or segments, several of these forming a complete ring in the
rotor or casing.

The method of forming the segments is illustrated by Fig. 387,
on Plate CLXVII., facing this page, which supplements the illustrations given on Plate XXVI., facing page 48. In the root ends of
the blades and packing pieces small holes, about $\frac{1}{8}$ in., are drilled for
stringing them on wires, which is done in cast-iron segmental pieces
or "formers," each representing a portion of the circumference of
the rotor or casing, and having the necessary blade grooves. The
"formers" are set up on a bench, as shown in Fig. 387. A piece
of brass wire of a gauge thickness just sufficient to pass through
the holes in the roots of the blade and packing piece is riveted
or soldered to one packing piece, which is then fixed into one end
of the groove of the cast-iron "former." The wire lies along
the path of the groove, and alternate blade pieces and packing

2 H

pieces are threaded on to the wire and are knocked up into the groove. This is continued until the requisite length of segment is obtained (usually about 30 in.), when the wire is riveted over the last packing piece. Should the segment require to have the outer tip fitted with a binding or shrouding wire, the operation of brazing this in the notches provided for the purpose and binding it is also performed at this stage. Silver solder is used as the brazing material.

The cast-iron "formers" are in halves, the abutting faces being along the line of the blade groove; it is therefore only necessary to slacken the bolts holding the two parts together to free the blade segment from the "former." These segments are generally stored until the respective casing or rotor is ready to receive them. In this way blading can proceed simultaneously with the machining of the casing or rotor. The view on Plate CLXVIII., facing this page, shows a group of segments of blades for rotors and casings.

In the fitting out of a casing or rotor, the segments are lightly knocked into the grooves with wooden mallets, and the packing pieces between the blades are calked hard into the grooves by suitable tools. The amount of calking necessary to make a secure job varies according to the size of the blade; the smaller blades receive two or three blows with an ordinary hand hammer, while in the case of the large blades, two or three blows with a 7 lb. hammer are sufficient to ensure a good job. Fig. 389, on Plate CLXIX., facing page 235, shows a rotor being bladed.

The rotors, after being fitted with blades, are put into the lathe and a fine turning is taken over the tips of the blades, giving the necessary clearance, which, in most cases, is but a small fraction of an inch, being about .03 in. to .04 in. in a rotor of about 5 ft. to 6 ft. in diameter. Great care and accurate measurement are necessary to ensure that this operation is successfully performed. To obtain clean cutting on such thin material as these blade tips, it is found necessary to run at a high cutting speed with a quick feed and a small cut, the tool being specially formed and having a very sharp edge. A similar procedure is carried out with the blading of the casings, but in this case the turning up of the blades is performed on the boring machine.

Plate CLXVIII.—Segments of Blades for Rotors and Casings (Fig. 388).

Fig. 389. Blading a Rotor at the Turbinia Works.

Fig. 390. Testing Turbines at the Turbinia Works.

Plate CLXIX.— Finishing Turbines.

The casing and rotor, having been bladed and fine tipped in their respective machines, are now thoroughly cleaned with a high-pressure steam jet and then go forward to the erecting tables. The completed rotor is afterwards put on the knife edges for the final static balancing and made as accurate as possible by the addition of suitable balance weights. The rotor is then bedded into the bearings, which have been fitted into the turbine casing ends, and the whole machine is checked from end to end as to clearances, thrust alignment, &c., and the top part of the casing fitted. Steam and exhaust pipes are then laid on to the turbine, and, after being thoroughly warmed up, the turbine is run under steam and the balancing checked at all speeds of revolution up to about 25 per cent. in excess of that demanded in service conditions. After the balance and steady running are proved, the turbine is kept under steam a sufficient time to consolidate the bearings and to ensure that the adjusting block rings are thoroughly bedded on to the rotor collars. Fig. 390, on Plate CLXIX., facing this page, shows a turbine being tested under steam.

After being again fully stripped and cleaned out, the turbine is jointed up permanently, is given a full run under steam to maximum revolutions, and is then ready for shipment. Where the weight is too great for the crane available the final fitting together must be done in the position the turbine is ultimately to occupy.

The construction of bearings, gland and other rings, governors, valves, &c., is in accordance with the usual practice, and need not be described. We shall therefore pass to a brief notice of the Parsons Works for the manufacture of land and marine turbines.

THE HEATON WORKS OF
MESSRS. C. A. PARSONS AND COMPANY.

THE Heaton Works, which are situated about two miles to the east of Newcastle-on-Tyne, were established in 1890 for the manufacture of steam turbines, high-speed dynamos, and electrical accessories generally. At that time Mr. Parsons was precluded from manufacturing the parallel-flow turbine, and something like four years' almost continuous research work was undertaken at the establishment in order to evolve a satisfactory radial-flow turbine. This experimental work (which is reviewed in pages 34 to 44) yielded valuable information, which was in part utilised when, in 1894, Mr. Parsons recovered the rights in the invention of the parallel-flow turbine and his original high-speed dynamo.

During the course of this experimental work, however, manufacturing was in progress, the first turbines made being, of course, of the radial-flow type. Prior to the foundation of the works, the total power of Parsons steam turbines in use was not much over 5000 horse-power, the largest unit being 120 horse-power. So soon, however, as it was possible to enter upon the manufacture of the parallel-flow turbine a great change was effected, and within a year or two the total power of turbines in use had increased to nearly 100,000 horse-power, the largest unit being 500 horse-power; while now, after the lapse of about twenty years, the total power of turbines built for duty on land in all countries exceeds 3,000,000 horse-power, the largest unit being capable, in association with its generator, of maintaining an output of 10,000 kilowatts.

This advance in the application of the turbine was partly in consequence of the improving economy. When the works were started a steam consumption of 42 lb. per kilowatt-hour non-condensing was regarded as satisfactory. It was reduced to 27 lb. per kilowatt in the first condensing radial turbine. When the first parallel-flow condensing turbine was made at the Heaton Works the consumption fell to 25 lb., whereas to-day, as has been established in preceding chapters, the consumption is little over 13 lb. per kilowatt-hour. This increased economy is, of course, the result of the improvement in design and growth in size which we have described.

It will be readily understood that in consequence of the greater demands the works at Heaton have undergone expansion in area and development in producing facilities. Originally the area covered was about two acres, and the largest shop was only about 170 ft. long by 50 ft. wide, consisting of two bays, with the requisite machinery for forming the casings, rotors, blades, and other accessories of turbines. Now, however, the works have an area of eleven acres, with extensive new shops and equipment to enable the larger units to be produced with the accuracy and reliability essential to the efficiency of the turbine.

In the preceding chapter we have described the general process of manufacturing turbines, and the intention in this chapter is to give a description of the factory where the work is carried on. We may begin with the pattern shop, of which a general view is given on Plate CLXX., facing page 238. This, as will be appreciated from remarks we made in the previous chapter, is an important department. The shop is 145 ft. long by 40 ft. wide, and there is in it every type of modern wood-working machinery. All are electrically driven, the countershaft being placed under the floor level. Adjoining this building is a large store for patterns of standard parts.

The principal machine shop is a new building having a length of 400 ft., divided into three bays of 40 ft. each, one of which is devoted to machining operations, the second partly to machining and partly to fitting together and erecting the turbines, and the third entirely to erecting turbines and electrical work. All of these bays, one of which is illustrated in Fig. 392, on Plate CLXX., facing page 238, are commanded by electric cranes of from 10 to 40 tons capacity. Reference need only be made to some of the more important machines. For dealing with the casings there is a planer 11 ft. 1 in. in width between standards, designed to take a cut 24 ft. long. A planer of slightly smaller size accommodates a casing 6 ft. 6 in. wide, the travel of the table being 22 ft. For vertical work there is a wall planing machine, usually devoted to the finishing of bed-plate faces and magnet and stator ends, while a milling machine is utilised for claw couplings and such parts, as well as for key seatings in large spindles. This miller takes units of 4 ft. square between the standards, the table having a travel of 8 ft. 6 in. A vertical lathe or boring mill deals with castings 6 ft. in diameter, and has two tool holders on the cross slide.

For boring casings there are five large boring mills of the horizontal type, the largest taking cylinders 15 ft. in diameter by 30 ft. long. There are traversing bars and slides, and rising and falling tables fitted with cross slides to facilitate accuracy in setting the job. The cutting of the grooves for the reception of the blades in the casing is also done in this machine, a special boring bar and apparatus being used, which automatically feeds the grooving tool radially outwards until a sufficient depth of groove has been cut for the complete circumference of the casing. The depth of cut is indicated by automatic gear on the end of the boring bar.

There is a very fine collection of lathes for turning the rotors both before and after blading. The largest is a treble-geared machine, having 4-ft. centres and a bed 35 ft. long, with three automatic slide rests. The train of gear is pretty extensive, so that there can be a very wide range in speed. This machine is not only utilised for turning the drum and forming the grooves, but also for tipping the ends of the blades after they have been fitted, in order to ensure absolute accuracy in their length. As already explained, this work is done at a high speed with a very slight cut. Where steel blades are used, as in blowers, the tipping is done by a high-speed electrically-driven emery wheel mounted on the slide rest of the lathe, instead of by the ordinary cutting tool. Another lathe of nearly the same capacity is used for turning very large armatures, cast-steel rotors for revolving field alternators, and grooving large commutators. A screw-cutting lathe, of 24-in.

Fig. 391. The Pattern Shop.

Fig. 392. One of the Machine Shops.

Plate CLXX. — The Heaton Works.

Fig. 393. The Erecting Shop.

Fig. 394. The Testing Department.

Plate CLXXI. — The Heaton Works.

centres and 30-ft. bed, is found a very useful machine; while a turret lathe, utilised for heavy bearings, has an inclined turret taking six tools. A large chuck lathe is used exclusively for machining the special discs utilised for building up certain of the turbine and blower shafts. This lathe is arranged with a special headstock, to enable a uniform cutting speed to be automatically maintained on any diameter of work. These, it will be understood, are only a few of the important tools. All of the large machines are driven by motors, the planers being operated by the Vickers electric controlling gear with reversing motors. The smaller machines, however, are driven by two lines of overhead shafting, operated by 50 horse-power electric motors.

The setting out is done on a marking-off table, 24 ft. long by 10 ft. wide, situated in the machine bay. As a large part of the work is standardised the parts are received into the store, which is located at the end of the machine shop. All parts manufactured are inspected and tested by the gauging department, and are classified so as to be passed down conveniently to the erecting department, which occupies the adjoining bay.

The erecting department, of which a view is given on Plate CLXXI., facing this page, is devoted to the building up of the turbines as well as of the blowing engines and electrical machinery. The process of erecting the turbine has, however, been so fully described in the preceding chapter that it is not necessary to enter into any details here. As many as twelve turbines have been in process of erection simultaneously, and, as we have already said, these have ranged up to machines capable of maintaining 10,000 kilowatts output.

The large units of the electrical dynamos, alternators, &c., are machined in the same shops as the parts of the turbines, and the small parts are manufactured in a separate building. This shop is equipped with complete plant for punching laminated discs for both continuous-current and alternating-current machines. The plates are all papered on one side in a special machine, and carefully scrubbed to remove the arrises caused by the punching and shearing processes. There are also vertical and horizontal winding machines, the latter for forming the compensating windings fitted to all continuous-current dynamos, as referred to in our chapter on "Electric Generators and the Turbines for Driving them."

The revolving units are all balanced separately before assembly, so that when the dynamo is complete practically nothing requires to be done in this direction, the revolving parts being often run up to full speed without any further balancing. The balancing house is a separate building erected of ferro-concrete, great care being taken to ensure stability, so that there can be no movement due to vibration when the turbine shaft and armature are run at their maximum speed for balancing purposes. Two adjustable bearings, lubricated with oil under pressure, are provided for taking the parts to be balanced. The driving pulley, fixed to the shaft, is rotated by belt from a 100 horse-power motor, which can be run at any speed from 500 to 800 revolutions per minute.

Among the many auxiliary departments — which include a blacksmith's shop, a tool room, extensive laboratories and experimental shops, with large stores—there is the mirror-constructing shop, which is of six bays, 30 ft. wide and 76 ft. long. In this the Parsons searchlight reflectors have been manufactured for over twenty-two years, the first orders having been for reflectors for use in the Suez Canal. These had a hyperbola and a parabola axis to give a flat beam ; later on parabola mirrors for searchlights have been manufactured, and are now very extensively used in the British and other naval services. The standard manufactures are 36 in. and 24 in. in diameter, but the equipment is such that reflectors of 7 ft. can be manufactured, and some of 5 ft. diameter have been already produced for the British Navy. A special feature is the parabola ellipse mirror, in which the beam from the mirror is concentrated on a narrow slit and then diverges out beyond, so that the projector can be placed behind a narrow loophole which only offers a small target for shots, whilst at the same time the whole of the light passes in the desired direction. This manufacture, however, is only mentioned to show the variety of the work done, and we may return to the main subject of steam turbines and deal with the important auxiliary of the testing department.

Before being despatched every turbo-generator is subjected to exhaustive trials, in order to ascertain whether the designed conditions have been realised. The results are very carefully collated and analysed, in order to win from them suggestions for future improvements. The testing house is a large building, 200 ft. long by 50 ft. wide, with a height from the floor to the eaves of 38 ft., so that the turbines and all the condensing plant may be accommodated in exactly the same relative positions as anticipated in the design. The work of placing the various units in position is carried out by two electric travelling cranes, each capable of lifting 20 tons. The floor, too, is of concrete, slightly sloped to a gully at one side for drainage purposes. There are permanent steam connections, with portable junction pieces and overhead electric wires for carrying off the current generated during the trial. A view of the testing shop is given in Fig. 394, on Plate CLXXI., facing page 239.

In view of the large units tested extensive steam-generating plant is necessary, the total capacity being 56,000 lb. of steam per hour. These boilers vary in type, including Lancashire, Babcock and Wilcox, Stirling, and Woodeson boilers. The last three (water-tube boilers) are worked at 350-lb. pressure, and the first at 200 lb. In view of the extent to which superheated steam is now utilised, there is fitted a separately-fired Babcock and Wilcox superheater, which can supply steam of 100 deg. Fahr. superheat at the rate of about 46,000 lb. per hour. Three chimneys are provided, 80 ft. high, and two are fitted with the Sturtevant system of induced draught.

The condensing plant is of the Parsons vacuum-augmenter type, with a total cooling surface of 4500 square feet. The circulating pumps are of the centrifugal type, and the air pump of Parsons vertical single-acting compound type. A vacuum of 28 in. to 29 in. is easily obtained with a barometrical pressure of 30 in. Water-measuring tanks are provided in connection with the testing of the turbines.

The method adopted in testing the electric machines is for the turbine spindle to be run up to its maximum speed, usually

25 per cent. above the normal, in order to ascertain whether the conditions of mechanical strength and true balance have been fulfilled. The dynamo rotor is then coupled up and is similarly tested, after which the open and short circuit curves of the dynamo are taken and the machine finally put on full load. In the later test the temperature of the dynamo is taken as well as the steam consumption of the turbine. As a rule, the turbine is run with a degree of superheat largely in excess of the ordinary working temperatures. When subjecting the dynamo to a high degree of overload the governing and voltage regulators are carefully tested. Alternators are also tried to ensure that they will conform to the same power factor as that of the system on which they are intended to run, and for this purpose choking coils are inserted in the circuits, so that any desired power factor can be obtained by regulating the number of turns of cable in the coils. The machinery is then opened out for inspection, and only if everything is satisfactory is the machine despatched to its destination.

The measuring apparatus in the testing department includes Kelvin balances and astatic watt-meters for enabling pressures up to 12,000 volts or 4000 ampere-units to be recorded. There is also a large oscillograph of the Duddell type, by which photographs of the voltage and current curves are produced under all conditions. Cinematograph records are also taken, by which very suggestive results have been obtained relative, for instance, to the effect on the load of switching on and off the feeders. There are two arrangements provided for putting on an artificial load by means of independent plates or wires submerged in water, one for low-voltage machines and the other for high-tension work. The former consists of a concrete tank into which the plates are suspended, while a constant supply of water flows over them, the level of the plates being varied in order to adjust the load. For high-tension work there is a similar contrivance, but, in order to ensure safety, the plates are placed in the centre of the cooling pond. The pond has an area of 32,000 square feet, and a capacity of 32,000 gallons, and is fitted with a Harrison set of spraying nozzles.

It will thus be seen that there is at the works a happy

combination of scientific investigation and sound and economical manufacture. This is as it should be, because it is not sufficient for the maintenance of a high reputation to depend exclusively upon past records. The aim should always be, as it is at the Heaton Works, to test every successive production, not only to determine whether it meets the stipulated condition, but also to ascertain if, and on what lines, improvements can be made in design and manufacture in order to increase efficiency.

PARSONS MARINE TURBINE WORKS AT WALLSEND.

THE Marine Turbine Works at Wallsend-on-Tyne were organised in 1897—thirteen years after the first turbine was manufactured for driving electric generators. The first patent specification had indicated the principles which should govern the application of the turbine for the propulsion of ships, and the second patent had established claims for a master patent for this application. Every promise of success, too, had been fulfilled in the results of the many turbo-electric generators made in the immediately succeeding years. And yet marine engineers, and, to a greater extent, shipowners, were somewhat conservative in their attitude towards the turbine. There was, perhaps, some justification for this in the high degree of economy attained by triple and quadruple-expansion reciprocating engines.

As has been explained fully in the introductory chapter, the practical demonstration, which alone was necessary to establish the potentialities of the marine turbine, was rendered possible by the formation in 1894 of a syndicate, for whom the first turbine steamer—the "Turbinia"—was constructed. The success attained, as described on pages 69 to 78, encouraged the further development of the system commercially under the same auspices. A new company—The Parsons Marine Steam Turbine Company, Limited— with wider scope and larger capital, was formed in 1897. This company established the marine works at Wallsend-on-Tyne, at which were designed and manufactured all marine turbines made up to the year 1904.

Since that date there have also been produced at Wallsend many of the notable engines of the type for the naval and merchant fleets, and a lead has thus been given in the chief advances in design, to the advantage of the numerous licensees under the patents.

This establishment, appropriately called the Turbinia Works, is situated four miles from Newcastle, on the north bank of the River Tyne, adjacent to the electric riverside lines of the North-Eastern Railway, from which there are sidings into the works. The area embraced is 23 acres, and the river frontage is 900 ft., with a wharf for the mooring of vessels during the period of fitting machinery on board, and for the despatch of turbines and machinery to other yards. To facilitate the work of placing the engines on board ship there are powerful sheerlegs and an electric jib crane on the wharf.

The workshops include machine and erecting shops, blading shops, test house, pattern shop, copper smithy, brass foundry, smithy, extensive stores, and an experimental department. Having been erected in recent years by Sir William Arrol and Co., Limited, the buildings are modern in every respect, with brick walls and steel columns, carrying a completely glazed roof on steel principals. The effect in respect of lighting, &c., is well illustrated by the engravings of three typical departments on Plates CLXXII., and CLXXIII., facing pages 246 and 247 respectively.

The original building scheme was conceived with due regard to extension as the potentialities of the turbine became established, so that additions have been easily made without affecting the convenience or symmetry of the plan. The successive changes indicate the development in the number and size of turbines manufactured. The two bays of the main shop, erected in 1897, had a height to the overhead crane track of 35 ft., and the largest crane had a capacity of 40 tons. When a third bay was erected in 1906 the height was made 45 ft., to suit the movement of larger rotors and casings, and the crane fitted was of 60 tons lifting power. In 1908 a second crane of similar capacity was added in the same erecting shop, so that when they are used together on one lift a load of 120 tons can be taken.

Similarly, while larger and more powerful tools have been added, there has been little need to depart in any essential particular from the design of machine tools originally devised by the Company for the special work of turbine manufacture. The improvements have been in detail, in order to ensure still greater accuracy and reliability in the finished result, rapidity in manufacture, and greater economy in production.

Electric power is, of course, used throughout the works, and it was only fitting that the current should be taken from the adjacent Carville Power Station, where the Parsons turbo-generator is extensively applied and efficiently run. There is in the works a large steam-producing plant, including one cylindrical and two water-tube boilers for use in testing the turbines before leaving the works.

In giving some idea of the equipment of the various departments we may begin with the pattern shop, which is illustrated on Plate CLXXII., facing this page. Of the practice adopted we have given some notes on page 228, and need only say that the building, which is 150 ft. long and 40 ft. wide, has a complete set of wood-working tools, driven by two electric motors with a counter-shaft under the floor level. The tools include two planing machines, circular, universal, and band saws, spindle machine, mortice and drilling machine, four lathes, a comprehensive wood-working machine, sand-papering machine, and hand trimmers. Benches with special vices have been arranged for about one hundred pattern-makers. Equable temperature is ensured by a system of hot-water pipes. Adjoining, there is a store about 100 ft. long and 40 ft. wide, built entirely of corrugated iron as a prevention against fire.

The brass foundry, also illustrated on the Plate facing this page, is 150 ft. long and 40 ft. wide, and is served by a 10-ton travelling crane. There are twelve fires, with a tall chimney to give a good natural draught. The drying stove, seen at the end of the shop as illustrated, has recently been rebuilt, and is 25 ft. long by 15 ft. wide by 15 ft. high. There are also to be seen along one side of the foundry a bench, with boxes for sand to facilitate the casting of small parts.

In the copper shop, which is also 150 ft. long and 40 ft. wide, there are nine brazing forges supplied with blast from a Root's blower. Compressed air pipes are laid throughout the shop in

Fig. 395. The Pattern Shop.

Fig. 396. The Foundry.

Plate CLXXII.— The Turbinia Works.

Plate CLXXIII.—The Machine Shop at the Turbinia Works (Fig. 397)

connection with portable pneumatic calking and riveting machines. The other appliances include a pneumatic hammer, a shearing and punching machine, an hydraulic pipe-bending machine, circular saws, and a drilling machine. A 5-ton travelling crane serves the part of the shop where the heavier items are worked, and in addition jib cranes swing over all the fires.

In the smithy, of the same size as the two shops just described, forgings up to 3 tons are produced with the use of special furnaces and a large power-driven pneumatic hammer. There are also a number of smith's hearths with water-cooled tuyeres for finishing small work.

The units when formed may be stored for use in a large separate building, where also the raw materials are housed until required, or may be passed at once to the machine shop to be turned, milled, planed, or otherwise formed to gauge, form, and dimensions

A view of one of the three bays of the machine and erecting shop is given on Plate CLXXIII , facing this page. Each bay is 360 ft. long and 45 ft. wide. The general procedure adopted in making casings and rotors has been described in a preceding chapter (pages 228 to 235), and here we confine our attention to a notice of the more important tools accommodated in the three bays of the shop. The largest boring machine will take casings up to 16 ft. in diameter and 50 ft. in length, but is arranged so that it can easily be altered to deal with a unit of larger dimensions. The most important lathe for turning rotors, also driven by a variable-speed motor, is one of the largest yet made. The planing of large cylinder casings is done on a heavy vertical and horizontal planer, seen to the left of the engraving.

In view of the success of the use of reduction gear, as initiated in the S.S. " Vespasian," large helical gear cutting machines have been installed; one of these is capable of dealing with gear wheels up to 13 ft. in diameter.

Beginning at the north end of the east bay there is, first, a 90-in. centre duplex sliding, surfacing, and screw-cutting lathe to take a job 55 ft. 6 in. between centres, and to swing 15 ft. over the bed and sliding carriages. This lathe is driven by a variable-speed direct-current motor. The fast headstock is fitted with powerful single, double, and treble gearing. There are four-speed cone pulleys, 27 in. to 36 in. in diameter by 7½ in. wide. A face-plate chuck, 12 ft. diameter, with external gearing, is provided, with four loose steel jaws on the front. There are also two independent sliding carriages, one at the front and the other at the back.

Next to this lathe is a horizontal boring machine, designed especially for boring out and grooving turbine casings. The machine is arranged to bore casings up to 16 ft. in diameter and 40 ft. in length. It has a massive driving headstock, with hollow spindle running in gun-metal adjustable bearings, and rotated by a machine-cut worm wheel and an adjustable worm. The latter has ball-thrust bearings at each end. The boring machine also is driven by a variable-speed motor, working through three changes of machine-cut gearing. This machine is a copy of one in the west bay, which the Parsons Company made when they instituted the works, and both are capable of taking off 2 in. on the diameter in one cut by means of six tools on one boring disc.

Beyond this boring machine is a new vertical and horizontal planing machine. The area covered by the tool is 20 ft. by 18 ft. The base plate upon which the work is fixed is 19 ft. by 7 ft. 6 in., and the machine is driven by a constant-speed motor. Next in order is a double horizontal drilling, boring, tapping, milling, and studding machine. The spindles are 4¾ in. in diameter, and the bed gives a 20-ft. travel horizontally. There are also in this bay two shafting lathes and a machine for calibrating torsion meters for measuring shaft horse-power.

There are several more large machines in the western bay. A boring mill, built by the Parsons Company, takes work up to 12 ft. in diameter, and was made specially for the manufacture of the turbines of the Allan liner "Virginian" in 1905. All the turbines of the British battleship "Dreadnought" were bored on this machine. There is a second boring mill to take casings up to 6 ft. in diameter. The casings for the British cruiser "Amethyst" were bored on this machine. There are two wall planers respectively to take cuts 10 ft. by 10 ft. and 12 ft. by 15 ft., two chuck lathes respectively with 3 ft. and 4 ft. centres, one lathe 4 ft. centre, a large double horizontal milling and boring machine, one large and several small horizontal planing machines, two slotters and six radial drills. Two of the latter are arranged on one bedplate, so that small turbine casings can be drilled simultaneously.

In the centre bay are the small lathes, including several turret and universal automatic tools, while here also are the fitting benches of the brass finishers. While all the large tools in the other bays are separately driven, a line of shafting is fitted for these small machines, and is rotated by a 30 horse-power three-phase motor. In this centre bay also are the tool shop and tool store, both self-contained and enclosed by wire netting. In the tool shop are benches and vices, as well as a screw-cutting lathe, a milling machine, several grinding machines, a large twist drill grinder to finish drills up to 4 in. in diameter, and several small drilling machines. Above the tool store is the foreman's office, and beyond it the bearing bench.

The southern end of all three bays is devoted to the erecting of rotors and casings, the blading of them, and the water and steam testing of the completed turbines. For the last-named process special foundations have been put down. This work was described in an earlier chapter (page 235).

There are two shops devoted entirely to the work of forming the blades to the correct size and to the making of segments of blades, as described in the chapter on the manufacture of turbines.

The first of these is a shop, 80 ft. long by 40 ft. wide, in which the blades are cut to the correct lengths by a Taylor and Challen patent blade press. The tips are thinned in the blade-tipping machines manufactured by the Parsons Company themselves, and the holes in the roots of the blades are drilled in small special drilling machines, also of the company's own manufacture. These also drill the holes through the sections after they have been cut to the correct lengths by a section-sawing machine. This work is illustrated on Plate CLXVII., facing page 232.

The blading material, when ready for assembling, is stored in boxes which have labels stating exactly for which expansion of each particular turbine the blades or sections are intended. When required, these boxes are taken to the second blade shop, where the assembling takes place.

This second blade shop is about 70 ft. long by 20 ft. wide, and has three lines of 3-in. by 3-in. angle irons, fixed about 3 ft. from the ground level, running the whole length of the shop. To these angle irons are attached the "formers" already described (page 49 *ante*), in which the blade segments on the "wire-root fixing" principle are made. An electric light is hung over every former, and gas is laid on at convenient places for use in brazing the binding strips. Outside this shop there is a lead-covered tank which holds a diluted solution of sulphuric acid, in which the segments, when completed, are immersed to be cleaned.

A special test house, 80 ft. long by 40 ft. wide and 35 ft. high, is used chiefly for steaming and testing the smaller sized turbines. A 30-ton travelling crane runs the whole length of the shop. There are two ranges of high-pressure steam pipes with numerous valves and down comers for testing purposes. At one end of this building, enclosed in a corrugated iron and glass lean-to, are the three-phase and direct-current distribution boards and a 100-kilowatt Vickers motor generator. For the last twelve months all the current for the works has been supplied, as already stated, by the Newcastle-upon-Tyne Electric Supply Company, the shafting, &c., being driven by 440-volt alternating-current motors direct, while the current for the cranes and for lighting is transformed to 110-volt direct current by a motor generator. Originally, however, energy was supplied by three of Messrs. C. A. Parsons and Co.'s turbo-dynamos which are still in position in the test house and can be coupled up in a few minutes to the direct-current switchboard in the event of a breakdown at the Carville Power Station.

Finally we come to the experimental department of the Turbinia Works, which has been an important factor in promoting the rapid development of the marine turbine and in the attainment of its high efficiency.

In the earlier chapters we have reviewed the course of experiments on turbine design carried on continuously from 1884 up to the present time, and have described how the confidence of ultimate success in the marine application of the turbine was founded on the results attained on land in the driving of dynamos. It has also been narrated that the "Turbinia" and her machinery were designed, and the turbines constructed, at the Heaton Works. From this stage onwards the marine development has been carried on at Wallsend, although the technical staffs of both establishments have been in close touch with each other, and the experimental resources at Heaton have been called upon in cases of difficulty. In many instances suggested improvements in turbines have been more cheaply and expeditiously investigated by trial on a turbine driving a dynamo where calibration of power and variation of speed are easy, and where a trial on board ship would have been too costly, if not impossible. Since the building of the "Turbinia" the chief advances may therefore be said to be the result of the concurrent work at both establishments, and both the marine and the land turbine have benefited by the experience and experimental research connected with each.

Many experiments in the details of design and construction, however, have necessarily devolved exclusively on the Turbinia Works, and with these we now have to deal. The chief characteristics of marine as compared with land turbine design are (1) that in the former there is set a high premium on the slowest possible rate of revolution and the least weight, consistent with the best economy in steam per shaft horse-power, and (2) that the limitations of space on board ship are in most cases of importance.

To meet these requirements in the best way has been the task of the technical staff at the Turbinia Works. This work has been carried on with much care and thoroughly exhaustive experimental investigation, and there has been no failure of a serious nature, notwithstanding the many great departures that have been made as the sizes of the engines increased year after year.

The work at Wallsend has been associated chiefly with the experimental testing and calculation of blading, blade strengths, and blade capacities in marine turbines. There is also apparatus for testing the efficiency of oil coolers, &c., so as to get the greatest effect in a given size of cooler. Recently, too,

there has been set up an experimental turbine and water brake
for trying various forms of impulse blading.

Important work has been done almost continuously in connection with the testing of propellers. The story of the development of the screws of the "Turbinia" is illustrative of the line on
which all such problems have been tackled. The results of these tests
on the "Turbinia," and also of later propeller tests, have advanced
our knowledge of propellers, regarding which there is, perhaps, less
exact information than of any other unit in marine machinery.

For high-speed ships a satisfactory over-all efficiency has been
attained by increasing, as far as conditions permit, the diameter
of the turbine to reduce the number of revolutions for a given
blade speed, and by modifying the design and proportions of the
propeller to work efficiently at a higher revolving rate than was
usual formerly.

Experiments in connection with propellers are being continued
on a larger and extended scale, and, with the great advantage of
extensive *data* obtained from the performance of existing ships,
the work of the past justifies expectations of still greater success
in the future.

The experimental department is also concerned with the form
of ships. Since 1900 model experiments have been in progress
in a tank over 100 ft. long, with a depth of water of 12 ft., and
arrangements have been made for accurately ascertaining the resistance of the models for given speeds and with various powers, while
wave lines can also be studied.

For the slow-speed steamer much good is likely to accrue from
the interposition between the turbine and the propeller of helical
gear, and this has been brought to high efficiency by a practical
test in the "Vespasian;" it has been found that the loss in
transmission is almost negligible—less than 2 per cent.

Thus the efficiency of machinery, whether for land or marine
purposes, can be tested in the experimental department of the
Heaton or the Turbinia Works from the generation of the steam
to its conversion to work in any form, and each unit can be studied
separately in order to determine any source of loss.

The record which we have given demonstrates the possession by the firm and its staff of a complete knowledge of the conditions of the problems to be solved, the accumulation of data and experience for guidance, and the necessary experimental skill and enterprise, as well as the faculty of concentration and perseverence, which win success.

Thus, although the story of the evolution of the Parsons turbine and of the subsequent developments is full in suggestiveness and in mechanical ingenuity, and although the turbine has been most beneficent in its influence on the economy of power production and in increase in speed in ships, there is justification for the hope that still better results will be reached in the future.

INDEX.

PRINTED AT THE BEDFORD PRESS, 20 AND 21, BEDFORDBURY, STRAND, LONDON, W.C.